Chemical

Discipline-Specific Review for the FE/EIT Exam

Second Edition

John R. Richards, PhD, PE
Stephanie T. Lopina, PhD, PE
with Michael R. Lindeburg, PE

The Power to Pass™
www.ppi2pass.com

Professional Publications, Inc. • Belmont, California

This product is part of PPI's Legacy Series. The Legacy Series contains products that were originally developed for the pencil-and-paper FE exam administered prior to 2014. PPI keeps these products in print to provide as thorough a selection of exam review options to our customers as possible. These products have not been revised to reflect the changes to the FE exam, but much of the content is appropriate for the current exam format, the books are still supported in PPI's errata management system, and the content is revised and updated to reflect errata submissions.

PPI continues to develop new products for the current computer-based FE exam, and some or all of the material in this book may appear in those newer products. For a complete list of available products, visit **ppi2pass.com/FE**.

Benefit by Registering This Book with PPI

- Get book updates and corrections.
- Hear the latest exam news.
- Obtain exclusive exam tips and strategies.
- Receive special discounts.

Register your book at **ppi2pass.com/register**.

Report Errors and View Corrections for This Book

PPI is grateful to every reader who notifies us of a possible error. Your feedback allows us to improve the quality and accuracy of our products. You can report errata and view corrections at **ppi2pass.com/errata**.

CHEMICAL DISCIPLINE-SPECIFIC REVIEW FOR THE FE/EIT EXAM
Second Edition

Current printing of this edition: 7

Printing History

date	edition number	printing number	update
Jan 2015	2	5	Minor corrections.
Feb 2015	2	6	Minor corrections.
Nov 2015	2	7	Minor corrections. Minor cover updates.

PPI
1250 Fifth Avenue
Belmont, CA 94002
(650) 593-9119
ppi2pass.com

Library of Congress Cataloging-in-Publication Data

Lopina, Stephanie T., 1964–
    Chemical discipline-specific review for the FE/EIT exam / Stephanie T. Lopina, with
    Michael R. Lindeburg.--2nd ed,
        p. cm.
    ISBN: 978-1-59126-067-7
    1. Chemical engineering--Problems, exercises, etc. 2. Chemical
engineering--Examinations, questions, etc. I. Lindeburg, Michael R. II. Title.
TP168.L66 2006
660.076--dc22
                                                            2005058665

F E D C B A

Table of Contents

Preface and Acknowledgments

This book is one in a series intended for engineers and students who are taking a discipline-specific (DS) afternoon session of the Fundamentals of Engineering (FE) exam.

The topics covered in the DS afternoon FE exams are completely different from the topics covered in the morning session of the FE exam. Since this book only covers one discipline-specific exam, it provides exam-level problems that are like those found on the afternoon half of the FE exam for the chemical discipline.

This book is intended to be a quick review of the material unique to the afternoon session of the chemical engineering exam. The material presented covers the subjects most likely to be on the exam. This book is not a thorough treatment of the exam topics. Its objective is to prepare you with enough knowledge to pass. As much as practical, this book uses the notation given in the NCEES Handbook.

This book consolidates 180 practical review problems, covering all of the discipline-specific exam topics. All problems include full solutions.

The first edition of this book was developed by Stephanie Lopina, PhD, PE, following the format, style, subject breakdown, and guidelines that I provided. John R. Richards wrote the original set of problems for Practice Exam 2. Kenneth Dewberry, PE, contributed significant updates of the practice problems and exam sections for the second edition.

In developing this book, the NCEES Handbook and the breakdown of problem types published by NCEES were my guide for problem types and scope of coverage. However, as with most standardized tests, there is no guarantee that any specific problem type will be encountered. It is expected that minor variations in problem content will occur from exam to exam.

As with all of PPI's books, the problems in this book are original and have been ethically derived. Although examinee feedback was used to determine its content, this book contains problems that are only *like* those that are on the exam. There are no actual exam problems in this book.

This book was designed to complement my *FE Review Manual*, which you will also need to prepare for the FE exam. The *FE Review Manual* is PPI's most popular study guide for this exam for more than 25 years.

You cannot prepare adequately without your own copy of the NCEES Handbook. This document contains the data and formulas that you will need to solve both the general and the discipline-specific problems. A good way to become familiar with it is to look up the information, formulas, and data that you need while trying to work practice problems.

Exam-prep books are always works in progress. By necessity, a book will change as the exam changes. Even when the exam format doesn't change for a while, new problems and improved explanations can always be added. I encourage you to provide comments via PPI's errata reporting page, **ppi2pass.com/errata**. You will find all verified errata there. I appreciate all feedback.

Best of luck to you in your pursuit of licensure.

Michael Lindeburg, PE

Engineering Registration in the United States

ENGINEERING REGISTRATION

Engineering registration (also known as *engineering licensing*) in the United States is an examination process by which a state's board of engineering licensing (i.e., registration board) determines and certifies that you have achieved a minimum level of competence. This process protects the public by preventing unqualified individuals from offering engineering services.

Most engineers do not need to be registered. In particular, most engineers who work for companies that design and manufacture products are exempt from the licensing requirement. This is known as the *industrial exemption.* Nevertheless, there are many good reasons for registering. For example, you cannot offer consulting engineering design services in any state unless you are registered in that state. Even within a product-oriented corporation, however, you may find that registered engineers have more opportunities for employment and advancement.

Once you have met the registration requirements, you will be allowed to use the titles Professional Engineer (PE), Registered Engineer (RE), and Consulting Engineer (CE).

Although the registration process is similar in all 50 states, each state has its own registration law. Unless you offer consulting engineering services in more than one state, however, you will not need to register in other states.

The U.S. Registration Procedure

To become a registered engineer in the United States, you will need to pass two eight-hour written examinations. The first is the *Fundamentals of Engineering Examination*, also known as the *Engineer-In-Training Examination* and the *Intern Engineer Exam.* The initials FE, EIT, and IE are also used. This exam covers basic subjects from the mathematics, physics, chemistry, and engineering classes you took during your first four university years. In rare cases, you may be allowed to skip this first exam.

The second eight-hour exam is the *Principles and Practice of Engineering Exam.* The initials PE are also used. This exam is on topics within a specific discipline, and only covers subjects that fall within that area of specialty.

Most states have similar registration procedures. However, the details of registration qualifications, experience requirements, minimum education levels, fees, oral interviews, and exam schedules vary from state to state. For more information, contact your state's registration board (**ppi2pass.com/stateboards**).

National Council of Examiners for Engineering and Surveying

The National Council of Examiners for Engineering and Surveying (NCEES) in Clemson, South Carolina, produces, distributes, and scores the national FE and PE exams. The individual states purchase the exams from NCEES and administer them themselves. NCEES does not distribute applications to take the exams, administer the exams or appeals, or notify you of the results. These tasks are all performed by the states.

Reciprocity Among States

With minor exceptions, having a license from one state will not permit you to practice engineering in another state. You must have a professional engineering license from each state in which you work. For most engineers, this is not a problem, but for some, it is. Luckily, it is not too difficult to get a license from every state you work in once you have a license from one state.

All states use the NCEES exams. If you take and pass the FE or PE exam in one state, your certificate will be honored by all of the other states. Although there may be other special requirements imposed by a state, it will not be necessary to retake the FE and PE exams. The issuance of an engineering license based on another state's license is known as *reciprocity* or *comity.*

The simultaneous administration of identical exams in all states has led to the term *uniform examination.* However, each state is still free to choose its own minimum passing score and to add special questions and requirements to the examination process. Therefore, the use of a uniform exam has not, by itself, ensured reciprocity among states.

THE FE EXAM

Applying for the Exam

Each state charges different fees, specifies different requirements, and uses different forms to apply for the exam. Therefore, it will be necessary to request an application from the state in which you want to become registered. Generally, it is sufficient for you to phone for this application. You'll find contact information (websites, telephone numbers, email addresses, etc.) for all U.S. state and territorial boards of registration at **ppi2pass.com/stateboards**.

Keep a copy of your exam application, and send the original application by certified mail, requesting a delivery receipt. Keep your proof of mailing and delivery with your copy of the application.

Exam Dates

The national FE and PE exams are administered twice a year (usually in mid-April and late October), on the same weekends in all states. For a current exam schedule, check **ppi2pass.com/fefaq**.

FE Exam Format

The NCEES Fundamentals of Engineering examination has the following format and characteristics.

- There are two four-hour sessions separated by a one-hour lunch.

- Examination questions are distributed in a bound examination booklet. A different exam booklet is used for each of the two sessions.

- Formulas and tables of data needed to solve questions in the exams are found in either the NCEES Handbook or in the body of the question statement itself.

- The morning session (also known as the *A.M. session*) has 120 multiple-choice questions, each with four possible answers lettered (A) through (D). Responses must be recorded with a pencil provided by NCEES on special answer sheets. No credit is given for answers recorded in ink.

- Each problem in the morning session is worth one point. The total score possible in the morning is 120 points. Guessing is valid; no points are subtracted for incorrect answers.

- There are questions on the exam from most of the undergraduate engineering degree program subjects. Questions from the same subject are all grouped together, and the subjects are labeled. The percentages of questions for each subject in the morning session are given in the following table.

Morning FE Exam Subjects

subject	percentage of questions (%)
chemistry	9
computers	7
electricity and magnetism	9
engineering economics	8
engineering probability and statistics	7
engineering mechanics (statics and dynamics)	10
ethics and business practices	7
fluid mechanics	7
material properties	7
mathematics	15
strength of materials	7
thermodynamics	7

- There are seven different versions of the afternoon session (also known as the *P.M. session*), six of which correspond to specific engineering disciplines: chemical, civil, electrical, environmental, industrial, and mechanical engineering.

The seventh version of the afternoon exam is a general examination suitable for anyone, but in particular, for engineers whose specialties are not one of the other six disciplines. Though the subjects in the general afternoon exam are similar to the morning subjects, the questions are more complex—hence their double weighting. Questions on the afternoon exam are intended to cover concepts learned in the last two years of a four-year degree program. Unlike morning questions, these questions may deal with more than one basic concept per question.

Each version of the afternoon session consists of 60 questions. All questions are mandatory. Questions in each subject may be grouped into related problem sets containing between two and ten questions each.

The percentages of questions for each subject in the general afternoon session exam are given in the following table.

Afternoon FE Exam Subjects (Other Disciplines Exam)

subject	percentage of questions (%)
advanced engineering mathematics	10
application of engineering mechanics	13
biology	5
electricity and magnetism	12
engineering economics	10
engineering probability and statistics	9
engineering of materials	11
fluids	15
thermodynamics and heat transfer	15

Each of the discipline-specific afternoon examinations covers a substantially different body of knowledge than the morning exam. The percentages of questions for each subject in the chemical discipline-specific afternoon session exam are as follows.

Afternoon FE Exam Subjects
(Chemical DS Exam)

subject	percentage of questions (%)
chemistry	10
material/energy balances	15
chemical engineering thermodynamics	10
fluid dynamics	10
heat transfer	10
mass transfer	10
chemical reaction engineering	10
process design and economic optimization	10
computer usage in chemical engineering	5
process control	5
safety, health, and environmental	5

Some afternoon questions stand alone, while others are grouped together, with a single problem statement that describes a situation followed by two or more questions about that situation. All questions are multiple choice. You must choose the best answer from among four, lettered (A) through (D).

- Each question in the afternoon is worth two points, making the total possible score 120 points.

- The scores from the morning and afternoon sessions are added together to determine your total score. No points are subtracted for guessing or incorrect answers. Both sessions are given equal weight. It is not necessary to achieve any minimum score on either the morning or afternoon sessions.

- All grading is done by computer optical sensing.

Use of SI Units on the FE Exam

Metric questions are used in all subjects, except some civil engineering and surveying subjects that typically use only customary U.S. (i.e., English) units. SI units are consistent with ANSI/IEEE standard 268 (the American Standard for Metric Practice). Non-SI metric units might still be used when common or where needed for consistency with tabulated data (e.g., use of bars in pressure measurement).

Grading and Scoring the FE Exam

The FE exam is not graded on the curve, and there is no guarantee that a certain percentage of examinees will pass. Rather, NCEES uses a modification of the Angoff procedure to determine the suggested passing score (the cutoff point or cut score).

With this method, a group of engineering professors and other experts estimate the fraction of minimally qualified engineers who will be able to answer each question correctly. The summation of the estimated fractions for all test questions becomes the passing score. Because the law in most states requires engineers to achieve a score of 70% to become licensed, you may be reported as having achieved a score of 70% if your raw score is greater than the passing score established by NCEES, regardless of the raw percentage. The actual score may be slightly more or slightly less than 110 as determined from the performance of all examinees on the equating subtest.

About 20% of the FE exam questions are repeated from previous exams—this is the *equating subtest*. Since the scores of previous examinees on the equating subtest are known, comparisons can be made between the two exams and examinee populations. These comparisons are used to adjust the passing score.

The individual states are free to adopt their own passing score, but all adopt NCEES's suggested passing score because the states believe this cutoff score can be defended if challenged.

You will receive the results within 12 weeks of taking the exam. If you pass, you will receive a letter stating that you have passed. If you fail, you will be notified that you failed and be provided with a diagnostic report.

Permitted Reference Material

Since October 1993, the FE exam has been what NCEES calls a "limited-reference" exam. This means that no books or references other than those supplied by NCEES may be used. Therefore, the FE exam is really an "NCEES-publication only" exam. NCEES provides its own Supplied-Reference Handbook for use during the examination. No books from other publishers may be used.

CALCULATORS

In most states, battery- and solar-powered, silent calculators can be used during the exam, although printers cannot be used. (Solar-powered calculators are preferable because they do not have batteries that run down.) In most states, programmable, preprogrammed, and business/finance calculators are allowed. Contact your state board to determine if nomographs and slide rules are permitted. To prevent unauthorized transcription and redistribution of the exam questions, calculators with communication or text-editing capabilities are banned from all NCEES exam sites. You cannot share calculators with other examinees. For a list of allowed calculators check **ppi2pass.com/calculators**.

It is essential that a calculator used for engineering examinations have the following functions.

- trigonometric functions
- inverse trigonometric functions
- hyperbolic functions
- pi
- square root and x^2
- common and natural logarithms
- y^x and e^x

For maximum speed, your calculator should also have or be programmed for the following functions.

- extracting roots of quadratic and higher-order equations
- converting between polar (phasor) and rectangular vectors
- finding standard deviations and variances
- calculating determinants of 3×3 matrices
- linear regression
- economic analysis and other financial functions

STRATEGIES FOR PASSING THE FE EXAM

The most successful strategy for passing the FE exam is to prepare in all of the exam subjects. Do not limit the number of subjects you study in hopes of finding enough questions in your strongest areas of knowledge to pass.

Fast recall and stamina are essential to doing well. You must be able to quickly recall solution procedures, formulas, and important data. You will not have time during the exam to derive solutions methods—you must know them instinctively. This ability must be maintained for eight hours. Be sure to gain familiarity with the NCEES Handbook by using it as your only reference for some of the problems you work during study sessions.

In order to get exposure to all exam subjects, it is imperative that you develop and adhere to a review schedule. If you are not taking a classroom review course (where the order of your preparation is determined by the lectures), prepare your own review schedule.

There are also physical demands on your body during the exam. It is very difficult to remain alert and attentive for eight hours or more. Unfortunately, the more time you study, the less time you have to maintain your physical condition. Thus, most examinees arrive at the exam site in peak mental condition but in deteriorated physical condition. While preparing for the FE exam is not the only good reason for embarking on a physical conditioning program, it can serve as a good incentive to get in shape.

It will be helpful to make a few simple decisions prior to starting your review. You should be aware of the different options available to you. For example, you should decide early on to

- use SI units in your preparation
- perform electrical calculations with effective (rms) or maximum values
- take calculations out to a maximum of four significant digits
- prepare in all exam subjects, not just your specialty areas

At the beginning of your review program, you should locate a spare calculator. It is not necessary to buy a spare if you can arrange to borrow one from a friend or the office. However, if possible, your primary and spare calculators should be identical. If your spare calculator is not identical to the primary calculator, spend some time familiarizing yourself with its functions.

A Few Days Before the Exam

There are a few things you should do a week or so before the exam date. For example, visit the exam site in order to find the building, parking areas, examination room, and rest rooms. You should also make arrangements now for child care and transportation. Since the exam does not always start or end at the designated times, make sure that your child care and transportation arrangements can tolerate a late completion.

Next in importance to your scholastic preparation is the preparation of your two examination kits. The first kit consists of a bag or box containing items to bring with you into the examination room.

[ ]   letter admitting you to the exam
[ ]   photographic identification
[ ]   main calculator
[ ]   spare calculator
[ ]   extra calculator batteries
[ ]   unobtrusive snacks
[ ]   travel pack of tissues
[ ]   headache remedy
[ ] $2.00 in change
[ ]   light, comfortable sweater
[ ]   loose shoes or slippers
[ ]   handkerchief
[ ]   cushion for your chair
[ ]   small hand towel
[ ]   earplugs
[ ]   wristwatch with alarm
[ ]   wire coat hanger
[ ]   extra set of car keys

The second kit consists of the following items and should be left in a separate bag or box in your car in case you need them.

[ ]   copy of your application
[ ]   proof of delivery
[ ]   this book
[ ]   other references
[ ]   regular dictionary
[ ]   scientific dictionary
[ ]   course notes in three-ring binders
[ ]   instruction booklets for all your calculators
[ ]   light lunch
[ ]   beverages in thermos and cans
[ ]   sunglasses
[ ]   extra pair of prescription glasses
[ ]   raincoat, boots, gloves, hat, and umbrella
[ ]   street map of the exam site
[ ]   note to the parking patrol for your windshield explaining where you are, what you are doing, and why your time may have expired
[ ]   battery-powered desk lamp

The Day Before the Exam

Take the day before the exam off from work to relax. Do not cram the last night. A good prior night's sleep is the best way to start the exam. If you live far from the exam site, consider getting a hotel room in which to spend the night.

Make sure your exam kits are packed and ready to go.

The Day of the Exam

You should arrive at least 30 minutes before the exam starts. This will allow time for finding a convenient parking place, bringing your materials to the exam room, and making room and seating changes. Be prepared, though, to find that the examination room is not open or ready at the designated time.

Once the examination has started, consider the following suggestions.

- Set your wristwatch alarm for five minutes before the end of each four-hour session, and use that remaining time to guess at all of the remaining unsolved problems. Do not work up until the very end. You will be successful with about 25% of your guesses, and these points will more than make up for the few points you might earn by working during the last five minutes.

- Do not spend more than two minutes per morning question. (The average time available per problem is two minutes.) If you have not finished a question in that time, make a note of it and move on.

- Do not ask your proctors technical questions. Even if they are knowledgeable in engineering, they will not be permitted to answer your questions.

- Make a quick mental note about any problems for which you cannot find a correct response or for which you believe there are two correct answers. Errors in the exam are rare, but they do occur. Being able to point out an error later might give you the margin you need to pass. Since such problems are almost always discovered during the scoring process and discounted from the exam, it is not necessary to tell your proctor, but be sure to mark the one best answer before moving on.

- Make sure all of your responses on the answer sheet are dark and completely fill the bubbles.

Common Questions About the DS Exam

Q: Do I have to take the DS exam?

A: Most people do not have to take the DS exam and may elect the general exam option. The state boards do not care which afternoon option you choose, nor do employers. In some cases, an examinee still in an undergraduate degree program may be required by his or her university to take a specific DS exam.

Q: Do all mechanical, civil, electrical, chemical, industrial, and environmental engineers take the DS exam?

A: Originally, the concept was that examinees from the "big five" disciplines would take the DS exam, and the general exam would be for everyone else. This remains just a concept, however. A majority of engineers in all of the disciplines apparently take the general exam.

Q: When do I elect to take the DS exam?

A: You will make your decision when registering for the FE exam.

Q: When do I choose which DS exam I want to take?

A: You must specify your desired DS exam when registering for the FE exam. Each topic is a separate booklet and you will receive only the exam indicated on your application.

Q: After I take the DS exam, does anyone know that I took it?

A: After you take the FE exam, only NCEES and your state board will know whether you took the DS or general exam. Such information may or may not be retained by your state board.

Q: Will my DS FE certificate be recognized by other states?

A: Yes. All states recognize passing the FE exam and do not distinguish between the DS and general afternoon portions of the FE exam.

Q: Is the DS FE certificate "better" than the general FE certificate?

A: There is no difference. No one will know which option you chose. It's not stated on the certificate you receive from your state.

Q: What is the format of the DS exam?

A: The DS exam is 4 hours long. There are 60 problems, each worth 2 points. The average time per problem is 4 minutes. Each problem is multiple choice with 4 answer choices. Most problems require the application of more than one concept (i.e., formula).

Q: Is there anything special about the way the DS exam is administered?

A: In all ways, the DS and general afternoon exam are equivalent. There is no penalty for guessing. No credit is given for scratch pad work, methods, etc.

Q: Are the answer choices close or tricky?

A: Answer choices are not particularly close together in value, so the number of significant digits is not going to be an issue. Wrong answers, referred to as "distractors" by NCEES, are credible. However, the exam is not "tricky"; it does not try to mislead you.

Q: Are any problems in the afternoon session related to each other?

A: Several questions may refer to the same situation or figure. However, NCEES has tried to make all of the questions independent. If you make a mistake on one question, it shouldn't carry over to another.

Q: Is there any minimum passing score for the DS exam?

A: No. It is the total score from your morning and afternoon sessions that determines your passing, not the individual session scores. You do not have to "pass" each session individually.

Q: Is the general portion easier, harder, or the same as the DS exams?

A: Theoretically, all of the afternoon options are the same. At least, that is the intent of offering the specific options—to reduce the variability. Individual passing rates, however, may still vary 5% to 10% from exam to exam. (PPI lists the most recent passing statistics for the various DS options on its website at **ppi2pass.com/fepassrates**.)

Q: Do the DS exams cover material at the undergraduate or graduate level?

A: Like the general exam, test topics come entirely from the typical undergraduate degree program. However, the emphasis is primarily on material from the third and fourth year of your program. This may put examinees who take the exam in their junior year at a disadvantage.

Q: Do you need practical work experience to take the DS exam?

A: No.

Q: Does the DS exam also draw on subjects that are in the general exam?

A: Yes. The dividing line between general and DS topics is often indistinct.

Q: Is the DS exam in customary U.S. or SI units?

A: The DS exam is nearly entirely in SI units. A few exceptions exist for some engineering subjects (surveying, hydrology, code-based design, etc.) where current common practice uses only customary U.S. units.

Q: Does the NCEES Handbook cover everything that is on the DS exam?

A: No. You may be tested on subjects that are not in the NCEES Handbook. However, NCEES has apparently adopted an unofficial policy of providing any necessary information, data, and formulas in the stem of the question. You will not be required to memorize any formulas.

Q: Is everything in the DS portion of the NCEES Handbook going to be on the exam?

A: Apparently, there is a fair amount of reference material that isn't needed for every exam. There is no way, however, to know in advance what material is needed.

Q: How long does it take to prepare for the DS exam?

A: Preparing for the DS exam is similar to preparing for a mini PE exam. Engineers typically take two to four months to complete a thorough review for the PE exam. However, examinees who are still in their degree program at a university probably aren't going to spend more than two weeks thinking about, worrying about, or preparing for the DS exam. They rely on their recent familiarity with the subject matter.

Q: If I take the DS exam and fail, do I have to take the DS exam the next time?

A: No. The examination process has no memory.

Q: Where can I get even more information about the DS exam?

A: If you have internet access, visit the Exam FAQs at **ppi2pass.com/fe.**

How to Use This Book

HOW EXAMINEES CAN USE THIS BOOK

This book is divided into three parts: The first part consists of 60 representative practice problems covering all of the topics in the afternoon DS exam. Sixty problems corresponds to the number of problems in the afternoon DS exam. You may time yourself by allowing approximately 4 minutes per problem when attempting to solve these problems, but that was not my intent when designing this book. Since the solution follows directly after each problem in this section, I intended for you to read through the problems, attempt to solve them on your own, become familiar with the support material in the official NCEES Handbook, and accumulate the reference materials you think you will need for additional study.

The second and third parts of this book consists of two complete sample examinations that you can use as sources of additional practice problems or as timed diagnostic tools. They also contain 60 problems, and the number of problems in each subject corresponds to the breakdown of subjects published by NCEES. Since the solutions to these parts of the book are consolidated at the end, it was my intent that you would solve these problems in a realistic mock-exam mode.

You should use the NCEES Handbook as your only reference during the mock exams.

The morning general exam and the afternoon DS exam essentially cover two different bodies of knowledge. It takes a lot of discipline to prepare for two standardized exams simultaneously. Because of that (and because of my good understanding of human nature), I suspect that you will be tempted to start preparing for your chosen DS exam only after you have become comfortable with the general subjects. That's actually quite logical, because if you run out of time, you will still have the general afternoon exam as a viable option.

If, however, you are limited in time to only two or three months of study, it will be quite difficult to do a thorough DS review if you wait until after you have finished your general review. With a limited amount of time, you really need to prepare for both exams in parallel.

HOW INSTRUCTORS CAN USE THIS BOOK

The availability of the discipline-specific FE exam has greatly complicated the lives of review course instructors and coordinators. The general consensus is that it is essentially impossible to do justice to all of the general FE exam topics and then present a credible review for each of the DS topics. Increases in course cost, expenses, course length, and instructor pools (among many other issues) all conspire to create quite a difficult situation.

One-day reviews for each DS subject are subject-overload from a reviewing examinee's standpoint. Efforts to shuffle FE students over the parallel PE review courses meet with scheduling conflicts. Another idea, that of lengthening lectures and providing more in-depth coverage of existing topics (e.g., covering transistors during the electricity lecture), is perceived as a misuse of time by a majority of the review course attendees. Is it any wonder that virtually every FE review course in the country has elected to only present reviews for the general afternoon exam?

But, while more than half of the examinees elect to take the other disciplines afternoon exam, some may actually be required to take a DS exam. This is particularly the case in some university environments where the FE exam has become useful as an "outcome assessment tool." Thus, some method of review is still needed.

Since most examinees begin reviewing approximately two to three months before the exam (which corresponds to when most review courses begin), it is impractical to wait until the end of the general review to start the DS review. The DS review must proceed in parallel with the general review.

In the absence of parallel DS lectures (something that isn't yet occurring in too many review courses), you may want to structure your review course to provide lectures only on the general subjects. Your DS review could be assigned as "independent study," using chapters and problems from this book. Thus, your DS review would consist of distributing this book with a schedule of assignments. Your instructional staff could still provide assistance on specific DS problems, and completed DS assignments could still be recorded.

The final chapter on incorporating DS subjects into review courses has yet to be written. Like the landscape architect who waits until a well-worn path appears through the plants before placing stepping stones, we need to see how review courses do it before we can give any advice.

Nomenclature

UNITS SYSTEMS

fundamental quantity	unit	abbreviation
SI Base Units		
length	meter	m
mass	kilogram	kg
time	second	s
temperature	kelvin	K
amount of substance	g-mole	mol
quantity of charge	coulomb	C
electric current	ampere	A
luminous intensity	candela	cd
volume	liter	L
U.S. Base Units		
length	foot	ft
mass	pound	lbm
time	hour	hr
temperature	rankine	°R
amount of substance	lb-mole	lbmol
luminous intensity	candle	cd
force	pound	lbf
energy	British thermal unit	Btu

NOMENCLATURE

A	annual amount	\$/yr
A	area	m^2
A	pre-exponential or frequency factor	–
$\hat{a}_i$	activity of component i in solution	–
A_{ij}	van Laar model constants for components i and j	–
B	molar bottoms product rate	mol/s
BP	boiling point	°C
c	velocity of sound	m/s
C	flow coefficient	–
C	number of velocity heads for fittings	–
C_A^*	equilibrium concentration	mol/m^3
Cap	capacity	–
C_i	concentration of component i	mol/m^3
C_{if}	feed concentration of component i	mol/m^3
C_{io}	feed concentration of component i	mol/m^3

c_p	specific heat capacity at constant pressure	J/kg·K
c_p	molar heat capacity at constant pressure	J/mol·K
c_p°	standard state heat capacity at constant pressure	J/mol·K
Δc_p°	molar heat capacity change at constant pressure	J/mol·K
c_v	specific heat capacity at constant volume	J/kg·K
c_v	molar heat capacity at constant volume	J/mol·K
d	distance	m
D	overhead product (distillate) molar flow rate	mol/s
D	diameter	m
D_m	diffusion coefficient	m^2/s
e	surface roughness	m
E_a	activation energy	J/mol
E_{ME}	Murphree plate efficiency	–
f	Darcy or Moody friction factor	–
F	feed molar flow rate	mol/s
F	feed moles in still pot	mol
F	future value	\$
F_A	moles of A flow rate	mol/s
F_{Ao}	moles of A fed per unit time	mol/s
f_i^L	fugacity of pure liquid component i	Pa
$\hat{f}_i^L$	fugacity of component i in the liquid phase	Pa
f_i°	fugacity of pure component i in its standard state	Pa
f_i^V	fugacity of pure vapor component i	Pa
$\hat{f}_i^V$	fugacity of component i in the vapor phase	Pa
g	local acceleration of gravity	m/s^2
g_c	Newton's law proportionality factor	kg·m/N·s^2
G	gas velocity	mol/m^2·h
ΔG°	standard Gibbs free energy of reaction	J/mol
ΔG_f^o	standard Gibbs free energy of formation	J/mol
h	height	m
h	individual heat-transfer coefficient	J/s·m^2·K

h	specific enthalpy	J/kg	K_p	pressure equilibrium constant	–
H	Henry's law constant		K_{sp}	solubility-product constant	–
	for concentration	Pa/(mol/m^3)	L	length	m
h_f	head loss due to pipe flow	m	L	liquid moles in still pot	mol
h_f	specific enthalpy of		L_{eq}	equivalent length fittings	m
	saturated liquid	J/kg	L_F	liquid molar flow rate	
$h_{f,\text{fitting}}$	head loss due to fittings	m		in feed	mol/s
$h_{f,\text{pump}}$	head due to pump work	m	L_n	liquid molar flow rate	
h_g	specific enthalpy of			leaving stage n	mol/s
	saturated vapor	J/kg	L_o	initial liquid moles	
Δh_r	heat of reaction	J/kg		in still pot	mol
ΔH	differential head	m	L_R	liquid molar flow rate	
$\Delta \hat{H}_c^\circ$	standard enthalpy change			in rectifying section	mol/s
	of combustion	J/mol	L_S	liquid molar flow rate	
$\Delta \hat{H}_f^\circ$	standard enthalpy change			in stripping section	mol/s
	of formation	J/mol	m	mass	kg
ΔH_r°	standard enthalpy change		m	molality	mol/kg
	of reaction	J	$\dot{m}$	mass flow rate	kg/s
k	reaction rate coefficient	–	$\dot{m}_e$	mass flow rate exit	kg/s
k	specific heat ratio	–	$\dot{m}_i$	mass flow rate inlet	kg/s
k	thermal conductivity	J/s·m·K	M	mach number	–
	for mole fraction	Pa	M	molarity	mol/L
K_a	activity equilibrium constant	–	M_i	molecular weight of	
K_b	boiling-point constant	°C/m		component i	kg/mol
ΔKE	kinetic energy change	J/kg	n	reaction rate order	–
K_{eq}	equilibrium constant	–	N	normality	eq/L
k_G	individual mass-transfer		N	number of moles	mol
	coefficient in gas for A		N	number of reactors	–
	diffusing through		N_A	molar flow of A	mol/s
	stagnant B	mol/s·m^2·Pa	NBP	normal boiling point	°C
k_G'	individual mass-transfer		N_I	diffusive flow of component I	mol/s
	coefficient in gas for		N_i	moles of component i	mol
	equimolar counter-diffusion	mol/s·m^2·Pa	N_{io}	feed moles of component i	mol
K_G	overall mass-transfer		Nu	Nusselt number	–
	coefficient in gas for		P	absolute pressure	Pa
	A diffusing through		P	power	W
	stagnant B	mol/s·m^2·Pa	P	present value	$
K_G'	overall mass-transfer		p_1	pressure at point 1	Pa
	coefficient in gas for		p_A^*	partial pressure in	
	equimolar counter-diffusion	mol/s·m^2·Pa		equilibrium with C_{AL}	Pa
k_i	Henry's law constant		$(p_B)_{lm}$	log mean of p_{B2} and p_{B1}	Pa
k_L	individual mass-transfer		p_c	critical pressure	Pa
	coefficient in liquid for		ΔPE	potential energy change	J/kg
	A diffusing through		p_i	partial pressure of	
	stagnant B	m/s		component i	Pa
k_L'	individual mass-transfer		P_i^{sat}	vapor pressure of pure	
	coefficient in liquid for			saturated i	Pa
	equimolar counter-diffusion	m/s	Pr	Prandtl number	–
K_L	overall mass-transfer		q	liquid feed condition,	
	coefficient in liquid			liquid in feed over feed	–
	for A diffusing through		Q	heat content	kcal/h
	stagnant B	m/s	Q	heat added to system	J
K_L'	overall mass-transfer		$\dot{Q}$	volumetric flow rate	m^3/s
	coefficient in liquid for		$\dot{Q}$	rate of heat addition	
	equimolar counter-diffusion	m/s		to system	J/s

$\dot{Q}_{in}$	rate of heat addition to system	J/s
R	universal gas constant	J/mol·K
$\bar{R}$	universal gas constant	J/mol·K
r_A	rate of reaction of component A	mol/m³·s
Ra_D	Rayleigh number	–
R_D	external reflux ratio	–
Re	Reynolds number	–
R_{fi}	fouling resistance inside	s-K/J
R_{fo}	fouling resistance outside	s-K/J
R_{min}	minimum external reflux ratio	–
s	specific entropy	J/kg·K
S	specific gravity	–
Sc	Schmidt number	–
s_f	specific entropy of saturated liquid	J/kg·K
s_g	specific entropy of saturated vapor	J/kg·K
Sh	Sherwood number	–
SV	space-velocity	s-1
t	thickness	m
t	time	s
T	absolute temperature	K
T_a	ambient temperature	K
ΔT_b	boiling-point elevation	°C
T_c	critical temperature	K
T_{Ci}	temperature cold-side inlet	K
T_{Co}	temperature cold-side outlet	K
T_{Hi}	temperature hot-side inlet	K
T_{Ho}	temperature hot-side outlet	K
ΔT_{lm}	log mean temperature difference	K
T_o	absolute reference temperature	K
T_o	temperature outside	K
T_P	absolute temperature of products	K
T_R	absolute temperature of reactants	K
T_∞	temperature at infinity	K
u	specific internal energy	J/kg
U	overall heat transfer coefficient	J/s·m²·K
u_g	specific internal energy of saturated vapor	J/kg
v	velocity	m/s
V	vapor moles in still pot	mol
V	volume	L
V	volume	m³
V_{CSTR}	volume of CSTR	m³
V_n	vapor molar flow rate leaving stage n	mol/s
v_o	volumetric flow rate	m³/s
V_o	initial volume	m³
V_{PFR}	volume of PFR	m³
V_R	vapor molar flow rate in rectifying section	mol/s
V_S	vapor molar flow rate in stripping section	mol/s
$V_{X_A=0}$	volume at 0% conversion	m³
$V_{X_A=1}$	volume at 100% conversion	m³
w	specific outward work done by the system	J/kg
W	outward work done by the system	J
W	weight in moles in still pot	mol
$\dot{W}$	rate of pump work (power)	W
$\dot{W}_{out}$	rate of outward work done by the system (power)	W
x	mole fraction of the more volatile component in liquid	–
X_i	conversion of component i	–
x_i	mole fraction of component i in the liquid	–
x_o	initial mole fraction of the more volatile component in liquid	–
y	mole fraction of the more volatile component in liquid	–
y_i	mole fraction of component i in the vapor	–
y_n	mole fraction of vapor leaving plate n	–
y_n^*	mole fraction of vapor in equilibrium with liquid on plate n	–
y_{n+1}	mole fraction of vapor leaving plate $n+1$	–
z	compressibility factor	–
z_1	elevation at point 1 above a datum level	m
z_F	mole fraction of the more volatile component in feed	–

SYMBOLS

α	relative volatility	–
β	volume expansivity	K^{-1}
γ	specific weight of liquid, ρg	kg/s²·m²
γ_i	activity coefficient of component i in the liquid	–

γ_i^∞	activity coefficient of component i at infinite dilution in the liquid	–
ε	emissivity	–
ε_A	fractional volume change on complete conversion	–
η	efficiency	–
μ	viscosity	cP
μ	viscosity	kg/m-s
ν	kinematic viscosity, μ/ρ	m²/s
ν_i	stoichiometric number of component i	–
ρ	density	g
ρ	specific density	kg/m³
σ	standard deviation	–
σ	Stefan-Boltzman constant	W/m² K4
τ	space-time	s
τ_i	space-time of ith reactor	s
v_i^L	specific volume of pure liquid i	m³/mol
v_f	specific volume of saturated liquid	m³/kg
v_g	specific volume of saturated vapor	m³/kg
$\hat{\Phi}_i$	fugacity coefficient of component i in the vapor	–
Φ_i^{sat}	fugacity coefficient of pure saturated i	–
Ω	reactor transverse area	m²

ME	Murphree efficiency
n	plate in rectifying section
o	outside, overhead, feedstate, initial state, or standard state
p	pipe or pressure
PFR	plug flow reactor
s	surface
S	stripping
sat	saturated
t	throat
v	volume
V	vapor
x	cross section
$*$	equilibrium
∞	infinity

SUBSCRIPTS AND SUPERSCRIPTS

1	point or state 1
2	point or state 2
a	activation or ambient
A	component A
avg	average
b	boiling point or bulk
B	bottoms product or component B
C	cold side
CSTR	continuous stirred tank reactor
D	overhead product (distillate)
e	exit
eq	equilibrium or equivalent
f	fluid, formation, or friction
F	feed
g	gas
G	gas
H	hot side
i	component i, inlet, or inside
L	liquid
lm	log mean
m	mercury or plate in stripping section

Practice Problems

CHEMICAL REACTION ENGINEERING

Problem 1

Compound A decomposes to compound R in a constant-volume isothermal batch reactor. The concentration of A is tracked over time. Using the data provided, determine the order of the irreversible stoichiometric reaction equation.

time (min)	C_A (mol/L)
0	20.00
2.45	11.21
5.21	7.50
8.36	5.44
12.72	3.94

(A) zeroth-order reaction
(B) first-order reaction
(C) second-order reaction
(D) third-order reaction

Solution

Rate equations are written for irreversible stoichiometric reactions of order 0 through 3. Integration of the rate equations yields relationships between reactant concentration and time that can be used to plot the experimental data.

order	reaction	rate equation	integrated form
0	$A \rightarrow R$	$-\dfrac{dC_A}{dt} = k$	$kt = C_{A0} - C_A$
1	$A \rightarrow R$	$-\dfrac{dC_A}{dt} = kC_A$	$kt = \ln\dfrac{C_{A0}}{C_A}$
2	$2A \rightarrow R$	$-\dfrac{dC_A}{dt} = kC_A^2$	$kt = \dfrac{1}{C_A} - \dfrac{1}{C_{A0}}$
3	$3A \rightarrow R$	$-\dfrac{dC_A}{dt} = kC_A^3$	$2kt = \dfrac{1}{C_A^2} - \dfrac{1}{C_{A0}^2}$

Manipulate the data to the relevant forms and plot the concentration term against time as in the following illustrations. The form for the correct reaction order will yield a straight line with slope k. Data from the second-order reaction yields a straight line.

time (min)	C_A (mol/L)	$C_{A0} - C_A$ (mol/L)	$\ln\dfrac{C_{A0}}{C_A}$	$\dfrac{1}{C_A} - \dfrac{1}{C_{A0}}$ (L/mol)	$\dfrac{1}{C_A^2} - \dfrac{1}{C_{A0}^2}$ (L/mol)^2
0.00	20.00	0.00	0.00	0.00	0.00
2.45	11.21	8.79	0.579	0.0392	0.00273
5.21	7.50	12.50	0.981	0.0833	0.00764
8.36	5.44	14.56	1.302	0.1338	0.01565
12.72	3.94	16.06	1.625	0.2038	0.03096

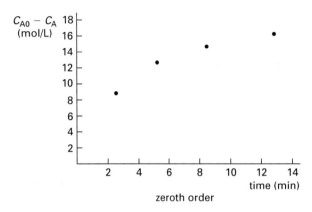

zeroth order

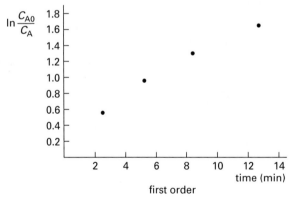

first order

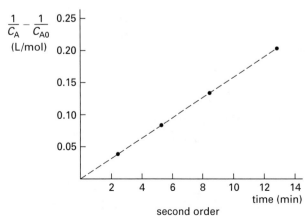

second order

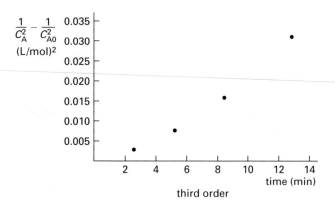

third order

The answer is C.

Problem 2

Under which condition does isothermal backmixing slow conversion to product?

- (A) reaction order -1
- (B) reaction order 0
- (C) reaction order 2
- (D) Isothermal backmixing does not slow conversion.

Solution

Compare the space-time, τ, in a continuously stirred tank reactor (CSTR) and a plug-flow reactor (PFR) for the given reaction orders. Space-time is the time required to process one reactor volume of feed.

$$\tau = \frac{C_{A0}V}{F_{A0}}$$

Complete backmixing is assumed in the CSTR, so it can be analyzed with a macroscopic material balance

$$\text{input} - \text{output} = \text{disappearance by reaction}$$
$$F_{A0} - F_{A0}(1 - X_A) = -r_A V$$
$$F_{A0}X_A = -r_A V$$
$$\tau_{\text{CSTR}} = \frac{C_{A0}V}{F_{A0}} = \frac{C_{A0}X_A}{-r_A}$$

No backmixing occurs in the PFR, which must be analyzed across a differential volume element.

$$\text{input} - \text{output} = \text{disappearance by reaction}$$
$$F_A - (F_A + dF_A) = -r_A dV$$
$$-d\big(F_{A0}(1 - X_A)\big) = -r_A dV$$
$$F_{A0}dX_A = -r_A dV$$
$$\tau_{\text{PFR}} = \frac{C_{A0}V}{F_{A0}}$$
$$= C_{A0}\int_0^{X_A} \frac{dX_A}{-r_A}$$

Reaction order -1:

$$-r = \frac{k}{C_A}$$
$$= \frac{k}{C_{A0}(1 - X_A)}$$
$$\tau_{\text{CSTR}} = C_{A0}X_A \frac{C_{A0}(1 - X_A)}{k}$$
$$= \frac{C_{A0}^2}{k} X_A(1 - X_A)$$
$$\tau_{\text{PFR}} = C_{A0} \int_0^{X_A} \frac{C_{A0}(1 - X_A)}{k} dX_A$$
$$= \frac{C_{A0}^2}{k} \left(X_A - \frac{X_A^2}{2} \right)$$
$$= \frac{C_{A0}^2}{2k} X_A(2 - X_A)$$
$$\frac{\tau_{\text{CSTR}}}{\tau_{\text{PFR}}} = \frac{2(1 - X_A)}{2 - X_A} < 1$$

Reaction order 0:

$$-r_A = k$$
$$\tau_{\text{CSTR}} = \frac{C_{A0}X_A}{k}$$
$$\tau_{\text{PFR}} = C_{A0} \int_0^{X_A} \frac{dX_A}{k} = \frac{C_{A0}X_A}{k}$$
$$\frac{\tau_{\text{CSTR}}}{\tau_{\text{PFR}}} = 1$$

Reaction order 2:

$$-r_A = kC_A^2$$
$$= kC_{A0}^2(1 - X_A)^2$$
$$\tau_{\text{CSTR}} = \frac{C_{A0}X_A}{kC_{A0}^2(1 - X_A)^2}$$
$$= \left(\frac{1}{kC_{A0}} \right) \left(\frac{X_A}{(1 - X_A)^2} \right)$$
$$\tau_{\text{PFR}} = C_{A0} \int_0^{X_A} \frac{dX_A}{kC_{A0}^2(1 - X_A)^2}$$
$$= \left(\frac{1}{kC_{A0}} \right) \left(\frac{1}{1 - X_A} - 1 \right)$$
$$= \left(\frac{1}{kC_{A0}} \right) \left(\frac{X_A}{1 - X_A} \right)$$
$$\frac{\tau_{\text{CSTR}}}{\tau_{\text{PFR}}} = \frac{1}{1 - X_A} > 1$$

For the second-order reaction, $\tau_{\text{CSTR}}/\tau_{\text{PFR}}$ is greater than one, which means more time is required to convert feed to product with backmixing.

The answer is C.

Problems 3–5 are based on the following information.

Component A decomposes in a first-order irreversible reaction to form R and S according to the following reaction in the liquid phase.

$$A(l) \rightarrow R(l) + S(g)$$

Pure A at a concentration of 31.7 mol/L is fed into a reactor system at 6.85 kmol/h and 140°C. The reaction rate constant at 140°C is 0.92 h^{-1}. Assume the following data are temperature independent.

heat of reaction of A $= -68.7$ kJ/kg
heat capacity of A $= 2.0$ kJ/kg·K
heat capacity of R $= 2.0$ kJ/kg·K
molecular weight of A $= 96.1$
molecular weight of R $= 80.1$
molecular weight of S $= 16.0$

Problem 3

If the reaction is carried out in an adiabatic PFR with a 30.0 cm inside diameter, the reactor length required to achieve 95% conversion is most nearly

(A) 0.0100 m
(B) 0.170 m
(C) 9.95 m
(D) 39.8 m

Solution

A material balance around a differential element yields

$$\text{input} - \text{output} = \text{disappearance by reaction}$$
$$F_A - (F_A + dF_A) = -r_A dV$$
$$dF_A = dF_{A0}(1 - X_A)$$
$$= -F_{A0} dX_A$$
$$F_{A0} dX_A = -r_A \Omega \, dz$$

The first-order reaction can be expressed as follows.

$$-r_A = kC_A$$
$$= kC_{A0}(1 - X_A)$$

Combining equations gives

$$\int_0^L dz = \frac{F_{A0}}{C_{A0} k \Omega} \int_0^{X_A} \frac{dX_A}{1 - X_A}$$

$$L = \frac{-F_{A0}}{C_{A0} k \Omega} \ln (1 - X_A)$$

$$= \left(\frac{-6850 \dfrac{\text{mol}}{\text{h}}}{\left(31.7 \dfrac{\text{mol}}{\text{L}} \right) (0.92 \text{ h}^{-1})} \times \left(\dfrac{\pi (0.30 \text{ m})^2}{4} \right) \left(1000 \dfrac{\text{L}}{\text{m}^3} \right) \right)$$
$$\times \ln (1 - 0.95)$$

$$= 9.95 \text{ m}$$

The answer is C.

Problem 4

What is the exit temperature from the PFR of Prob. 3? (Due to its low molecular weight, ignore component S.)

(A) 107°C
(B) 140°C
(C) 143°C
(D) 173°C

Solution

For an adiabatic system, the energy balance is

$$\begin{array}{c} \text{change of} \\ \text{heat content} \end{array} = \begin{array}{c} \text{heat effect of} \\ \text{chemical reactions} \end{array}$$

$$\sum_j \frac{F_j}{\Omega} c_{p_j} \frac{dT}{dz} = -r_A(-\Delta h_r)$$

Because $c_{p_A} = c_{p_R}$, the equation can be simplified. Substitute the expression for $-r_A$ derived for Prob. 3.

$$\frac{F}{\Omega} c_p \frac{dT}{dz} = \left(\frac{F_{A0}(-\Delta h_r)}{\Omega} \right) \left(\frac{dX_A}{dz} \right)$$

$$dT = \frac{F_{A0}(-\Delta h_r)}{F c_p} dX_A$$

The feed is pure A, so $F = F_{A0}$.

$$\int_{\text{in}}^{\text{out}} dT = \frac{-\Delta h_r}{c_p} \int_0^{X_A} dX_A$$

$$T_{\text{out}} - T_{\text{in}} = \frac{-\Delta h_r}{c_p} X_A$$

$$T_{\text{out}} = 140°C + \left(\frac{68.7 \dfrac{\text{kJ}}{\text{kg}}}{2.0 \dfrac{\text{kJ}}{\text{kg·K}}} \right) (0.95)$$

$$= 172.6°C \quad (173°C)$$

The answer is D.

Problem 5

If three identical CSTRs are run in series and the total conversion is 95%, the space-time for each reactor is most nearly

    (A) 0.688 h
    (B) 1.58 h
    (C) 1.86 h
    (D) 2.45 h

Solution

$$\tau = \frac{V}{v_0}$$

The space-time for each of three identical CSTRs in series will be the same; that is, $\tau_1 = \tau_2 = \tau_3 = \tau$.

Complete a mass balance on each of the reactors in series.

	entering R_1	entering R_2	entering R_3	exiting R_3
volumetric flow rate	V_1	$V_1 = V_0$	$V_2 = V_0$	$V_3 = V_0$
molar flow rate	F_{A0}	F_{A1}	F_{A2}	F_{A3}
concentration	C_{A0}	C_{A1}	C_{A2}	C_{A3}
mole fraction	x_{A0}	x_{A1}	x_{A2}	x_{A3}

Reactor 1: $V_0 C_{A0} - V_0 C_{A1} = VkC_{A1}$

$$C_{A1} = \frac{C_{A0}}{1 + k\tau_1}$$

Reactor 2: $V_0 C_{A1} - V_0 C_{A2} = VkC_{A2}$

$$C_{A2} = \frac{C_{A1}}{1 + k\tau_2}$$

Reactor 3: $V_0 C_{A2} - V_0 C_{A3} = VkC_{A3}$

$$C_{A3} = \frac{C_{A2}}{1 + k\tau_3}$$

Successive substitution yields

$$\frac{C_{A3}}{C_{A0}} = \frac{1}{(1 + k\tau)^3}$$

$$\tau = \frac{1}{k}\left(\left(\frac{C_{A0}}{C_{A3}}\right)^{1/3} - 1\right)$$

$$= \frac{1}{k}\left(\left(\frac{1}{1 - X_{A3}}\right)^{1/3} - 1\right)$$

$$= \left(\frac{1}{0.92 \text{ h}^{-1}}\right)\left(\left(\frac{1}{1 - 0.95}\right)^{1/3} - 1\right)$$

$$= 1.86 \text{ h}$$

The answer is C.

Problem 6

Consider the following reaction.

$$A \underset{k_2}{\overset{k_1}{\rightleftharpoons}} I_1 \underset{k_4}{\overset{k_3}{\rightleftharpoons}} I_2 \underset{k_6}{\overset{k_5}{\rightleftharpoons}} R$$

Assuming first-order kinetics, the rate of disappearance of I_1 can be expressed as

    (A) $-\dfrac{dC_{I_1}}{dt} = -\dfrac{k_1}{k_2}C_A + \dfrac{k_3}{k_4}C_{I_2} + \dfrac{k_5}{k_6}C_R$

    (B) $-\dfrac{dC_{I_1}}{dt} = -k_1 C_A + (k_2 + k_3)C_{I_1}$
                     $- (k_4 + k_5)C_{I_2} - k_6 C_R$

    (C) $-\dfrac{dC_{I_1}}{dt} = -\dfrac{k_1}{k_2}C_A + \dfrac{k_3}{k_4}C_{I_2}$

    (D) $-\dfrac{dC_{I_1}}{dt} = -k_1 C_A + (k_2 + k_3)C_{I_1} - k_4 C_{I_2}$

Solution

The reactions producing and consuming I_1 can be broken down as follows.

reaction	rate
$A \xrightarrow{k_1} I_1$	$-\dfrac{dC_{I_1}}{dt} = -k_1 C_A$
$I_1 \xrightarrow{k_2} A$	$-\dfrac{dC_{I_1}}{dt} = k_2 C_{I_1}$
$I_1 \xrightarrow{k_3} I_2$	$-\dfrac{dC_{I_1}}{dt} = k_3 C_{I_1}$
$I_2 \xrightarrow{k_4} I_1$	$-\dfrac{dC_{I_1}}{dt} = -k_4 C_{I_2}$

Summing the rate expressions gives the overall rate expression for I_1.

$$-\frac{dC_{I_1}}{dt} = -k_1 C_A + k_2 C_{I_1} + k_3 C_{I_1} - k_4 C_{I_2}$$
$$= -k_1 C_A + (k_2 + k_3)C_{I_1} - k_4 C_{I_2}$$

The answer is D.

CHEMICAL ENGINEERING THERMODYNAMICS

Problems 7–10 are based on the following information and illustration.

Consider the steam power plant shown. Assume that the condenser and boiler pressures are constant and that the turbines operate reversibly and adiabatically. Neglect any enthalpy change through the pump. Make-up water is added at the same temperature as the saturated liquid condensate.

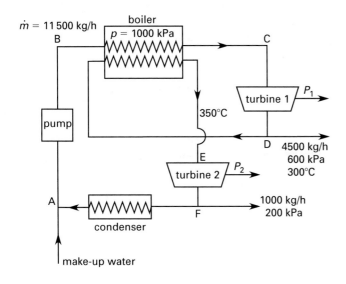

$\dot{m} = 11\,500$ kg/h

boiler
$p = 1000$ kPa

B

C

turbine 1 — P_1

350°C

pump

D   4500 kg/h
600 kPa
300°C

E

turbine 2 — P_2

A

1000 kg/h
200 kPa

F

condenser

make-up water

Problem 7

What is the total power developed by both turbines?

(A) 140 kW
(B) 280 kW
(C) 470 kW
(D) 990 kW

Solution

The mass flow rates are

$$\dot{m}_1 = 11\,500 \text{ kg/h}$$

$$\dot{m}_2 = 11\,500 \; \frac{\text{kg}}{\text{h}} - 4500 \; \frac{\text{kg}}{\text{h}}$$

$$= 7000 \text{ kg/h}$$

The steam leaving the first turbine is at 300°C and 600 kPa. The enthalpy and entropy can be found in the steam tables.

$$h_{\mathrm{D}} = 3061.6 \text{ kJ/kg}$$

$$s_{\mathrm{D}} = 7.3724 \text{ kJ/kg·K}$$

The entropy of the steam entering the turbine must be the same as that of the steam exiting the turbine, and the pressure is 1000 kPa. Again from the steam tables,

$$h_{\mathrm{C}} = 3203.9 \text{ kJ/kg}$$

$$T_{\mathrm{C}} = 371.74°\text{C}$$

The steam entering the second turbine is at 350°C and 600 kPa.

$$h_{\mathrm{E}} = 3165.7 \text{ kJ/kg}$$

$$s_{\mathrm{E}} = 7.5464 \text{ kJ/kg·K}$$

Exiting, the pressure is 200 kPa and the entropy remains constant. Interpolating from the superheated table,

$$T_{\mathrm{F}} \approx 210°\text{C}$$

$$h_{\mathrm{F}} = 2891 \text{ kJ/kg}$$

The power developed by adiabatic reversible turbines can be expressed as follows.

$$P = -\dot{m}_1(h_{\mathrm{D}} - h_{\mathrm{C}}) - \dot{m}_2(h_{\mathrm{F}} - h_{\mathrm{E}})$$

$$= \left(\begin{array}{c} \left(-11\,500 \; \frac{\text{kg}}{\text{h}} \right) \left(3061.6 \; \frac{\text{kJ}}{\text{kg}} - 3203.9 \; \frac{\text{kJ}}{\text{kg}} \right) \\ - \left(7000 \; \frac{\text{kg}}{\text{h}} \right) \left(2891 \; \frac{\text{kJ}}{\text{kg}} - 3165.7 \; \frac{\text{kJ}}{\text{kg}} \right) \end{array} \right)$$

$$\times \left(\frac{1 \text{ h}}{3600 \text{ s}} \right)$$

$$= 989 \text{ kW} \quad (990 \text{ kW})$$

The answer is D.

Problem 8

Calculate the heat requirement for the boiler.

(A) 1.2 MW
(B) 1.6 MW
(C) 8.7 MW
(D) 8.8 MW

Solution

At constant pressure, the heat required for the boiler is

$$\dot{Q} = \dot{m}_1(h_{\mathrm{C}} - h_{\mathrm{B}}) + \dot{m}_2(h_{\mathrm{E}} - h_{\mathrm{D}})$$

The problem statement indicates the enthalpy of saturated liquid is at 200 kPa and that $h_{\mathrm{B}} = h_{\mathrm{A}}$. From the steam tables,

$$h_{\mathrm{B}} = h_{\mathrm{A}} = 504.69 \text{ kJ/kg}$$

$$\dot{Q} = \left(\begin{array}{c} \left(11\,500 \; \frac{\text{kg}}{\text{h}} \right) \left(3203.9 \; \frac{\text{kJ}}{\text{kg}} - 504.69 \; \frac{\text{kJ}}{\text{kg}} \right) \\ + \left(7000 \; \frac{\text{kg}}{\text{h}} \right) \left(3165.7 \; \frac{\text{kJ}}{\text{kg}} - 3061.6 \; \frac{\text{kJ}}{\text{kg}} \right) \end{array} \right)$$

$$\times \left(\frac{1 \text{ h}}{3600 \text{ s}} \right)$$

$$= 8825 \text{ kW} \quad (8.8 \text{ MW})$$

The answer is D.

Problem 9

The power plant efficiency is most nearly

(A) 5.5%
(B) 11%
(C) 14%
(D) 40%

Solution
Power plant efficiency is the ratio of work produced to
heat supplied.

$$\eta = \frac{P}{Q} = \left(\frac{989 \text{ kW}}{8825 \text{ kW}}\right) \times 100\%$$

$$= 11.2\% \quad (11\%)$$

The answer is B.

Problem 10

Calculate the efficiency of a Carnot cycle operating between the boiler and condenser saturation temperatures.

   (A) 8.0%
   (B) 37%
   (C) 39%
   (D) 73%

Solution
The efficiency of a Carnot cycle can be calculated from
the hot and cold absolute temperatures in the cycle.

$$\eta_C = \frac{T_{\text{hot}} - T_{\text{cold}}}{T_{\text{hot}}}$$

The hot temperature is the boiler temperature, the temperature of stream C.

$$T_{\text{hot}} = 371.74°\text{C} + 273° = 644.74\text{K}$$

The cold temperature is the condenser temperature, the
temperature of saturated liquid at 200 kPa.

$$T_{\text{cold}} = 120.23°\text{C} + 273° = 393.23\text{K}$$

$$\eta_C = \left(\frac{644.74\text{K} - 393.23\text{K}}{644.74\text{K}}\right) \times 100\%$$

$$= 39.01\% \quad (39\%)$$

The answer is C.

Problems 11 and 12 are based on the following information.

Air at 22.0°C and 50.0% relative humidity passes
through a cooling tower at a rate of 2500 m³/min. The
saturated air exits the tower at 31.0°C. Water enters
the tower at 35.0°C and 3250 kg/min.

Problem 11

The water exits the tower at a temperature of

   (A) 21.4°C
   (B) 22.0°C
   (C) 23.1°C
   (D) 26.0°C

Solution
The enthalpies of the inlet and outlet air are read from
the psychrometric chart.

$$h_{\text{in}} = 43.1 \text{ kJ/kg}_{\text{dry air}}$$

$$h_{\text{out}} = 105.5 \text{ kJ/kg}_{\text{dry air}}$$

The specific volume of the inlet air stream is also read
from the psychrometric chart.

$$V = 0.847 \text{ m}^3/\text{kg}_{\text{dry air}}$$

The water temperature drop is calculated from the enthalpy change through the tower.

$$\dot{Q} = \dot{m}_w c_p (T_{\text{in}} - T_{\text{out}}) = \dot{m}_{\text{air}}(h_{\text{out}} - h_{\text{in}})$$

$$T_{\text{out}} = T_{\text{in}} - \frac{\dot{m}_{\text{air}}(h_{\text{out}} - h_{\text{in}})}{\dot{m}_w c_p}$$

$$= 35.0°\text{C}$$

$$- \frac{\left(\dfrac{2500 \ \dfrac{\text{m}^3}{\text{min}}}{0.847 \ \dfrac{\text{m}^3}{\text{kg}_{\text{dry air}}}}\right)\left(105.5 \ \dfrac{\text{kJ}}{\text{kg}_{\text{dry air}}} - 43.1 \ \dfrac{\text{kJ}}{\text{kg}_{\text{dry air}}}\right)}{\left(3250 \ \dfrac{\text{kg}}{\text{min}}\right)\left(4.18 \ \dfrac{\text{kJ}}{\text{kg}\cdot°\text{C}}\right)}$$

$$= 21.4°\text{C}$$

The answer is A.

Problem 12

How much make-up water is required?

   (A) 18 kg/min
   (B) 35 kg/min
   (C) 53 kg/min
   (D) 61 kg/min

Solution
The water content of the entering and exiting air can
be read from a psychrometric chart.

$$\omega_{\text{in}} = 8.25 \text{ g}_{\text{water}}/\text{kg}_{\text{dry air}}$$

$$\omega_{\text{out}} = 29.0 \text{ g}_{\text{water}}/\text{kg}_{\text{dry air}}$$

$$\Delta\omega = \omega_{\text{out}} - \omega_{\text{in}}$$

$$= 29.0 \ \frac{\text{g}_{\text{water}}}{\text{kg}_{\text{dry air}}} - 8.25 \ \frac{\text{g}_{\text{water}}}{\text{kg}_{\text{dry air}}}$$

$$= 20.75 \text{ g}_{\text{water}}/\text{kg}_{\text{dry air}}$$

The specific volume of the inlet air stream can be read
from the same psychrometric chart.

$$v = 0.847 \text{ m}^3/\text{kg}_{\text{dry air}}$$

The quantity of make-up water can be calculated as follows.

$$\text{make-up water} = (\text{moisture added})(\text{flow rate})$$

$$= \left(\frac{20.75 \, \frac{g_{water}}{kg_{dry\ air}}}{0.847 \, \frac{m^3}{kg_{dry\ air}}} \right) \left(\frac{2500 \, \frac{m^3}{min}}{1000 \, \frac{g}{kg}} \right)$$

$$= 61.2 \text{ kg/min} \quad (61 \text{ kg/min})$$

The answer is D.

COMPUTER USAGE IN CHEMICAL ENGINEERING

Problem 13

The cells in a spreadsheet application are defined as follows.

	A	B	C	D
1	3	12	=sum(C2:D4)	7
2	5	=average(C2:D4)	−1	=C3*2
3	=D2**2	=C1−B2	=A1+B4	−10
4	=ABS(B3)	−4	8	=min(A1:A3)

What is the value of cell A4?

(A) 0.83
(B) 1.0
(C) 2.5
(D) 8.3

Solution

Calculating the value of cell A4 requires the calculation of all of those cells that affect its result, directly or indirectly.

C3 = A1 + B4 = 3 − 4 = −1
D2 = C3*2 = −1*2 = −2
A3 = D2**2 = (−2)**2 = 4
D4 = min(A1:A3) = min(A1,A2,A3) = min(3,5,4) = 3
C1 = sum(C2:D4) = C2 + C3 + C4 + D2 + D3 + D4
$\quad$ = −1 − 1 + 8 − 2 − 10 + 3 = −3
B2 = average(C2:D4)
$\quad$ = (C2 + C3 + C4 + D2 + D3 + D4)/6 = −0.5
B3 = C1 − B2 = −3 − (−0.5) = −2.5
A4 = ABS(B3) = |−2.5| = 2.5

The answer is C.

Problem 14

Batch viscosity data from 10 production runs are entered into an array P. Values in units of Pa·s are 1.04, 0.85, 0.37, 1.73, 1.29, 0.94, 1.19, 1.09, 0.79, and 1.63.

Determine the output after the following pseudocode is executed.

```
Set A = 0.7; B = 1.3; N = 0; W = 0; X = 0;
    Y = 0; Z = 0;
LOOPSTART
    N = N + 1
    IF N > 10 THEN GO TO END
    IF P(N) < A THEN GO TO ONE
    IF P(N) > B THEN GO TO TWO
    Z = Z + 1
    W = W + P(N)
    GO TO LOOPSTART
ONE
    X = X + 1
    GO TO LOOPSTART
TWO
    Y = Y + 1
    GO TO LOOPSTART
END
    W = W / Z
    PRINT W,X,Y,Z
```

The output of the code will be

(A) 1.03, 1, 2, 7
(B) 1.03, 1, 3, 7
(C) 1.09, 1, 2, 7
(D) 1.09, 2, 3, 7

Solution

The pseudocode counts the number of batches where the viscosity is less than 0.7 Pa·s (X), the number of batches where the viscosity is greater than 1.3 Pa·s, (Y), and the number of batches where the viscosity is between 0.7 Pa·s and 1.3 Pa·s (Z). Then the program calculates the average viscosity for the Z batches.

$$X = 1$$
$$Y = 2$$
$$Z = 7$$

$$W = \frac{\begin{array}{c}1.04 + 0.85 + 1.29 + 0.94\\ + 1.19 + 1.09 + 0.79\end{array}}{7}$$
$$= 1.03$$

The output of the program is 1.03, 1, 2, 7.

The answer is A.

Problem 15

A nonideal gas is compressed from 24.8 m³ to 1.27 m³ in a reversible process as tabulated below.

V (m^3)	p (kPa)
24.8	101
20.1	199
15.4	405
10.7	655
5.98	1220
1.27	2130

Work done in the process can be expressed as follows.

$$W = \int p\,dV$$

Use the trapezoidal rule to calculate the work required to compress the gas. The answer is most nearly

(A) 14 MJ
(B) 17 MJ
(C) 22 MJ
(D) 25 MJ

Solution

The trapezoidal rule subdivides the area under the pV curve into a series of trapezoids and sums the areas.

$$W = \int p\,dV$$

$$\approx \frac{1}{2}\sum_{j=1}^{n-1}(V_{j+1} - V_j)(p_j + p_{j+1})$$

Since the volume measurements are equally spaced, any two consecutive steps can be used to represent ΔV, and the expression can be simplified as follows.

$$W = \int p\,dV$$

$$\approx \frac{\Delta V}{2}\left(p_1 + p_n + 2\sum_{j=2}^{n-1}p_j\right)$$

$$\approx \left(\frac{5.98\text{ m}^3 - 1.27\text{ m}^3}{2}\right)$$

$$\times \left(\begin{array}{l}101\text{ kPa} + 2130\text{ kPa} \\ + (2)\left(\begin{array}{l}199\text{ kPa} + 405\text{ kPa} \\ + 655\text{ kPa} + 1220\text{ kPa}\end{array}\right)\end{array}\right)$$

$$= 16\,930\text{ kJ}\quad(17\text{ MJ})$$

The answer is B.

HEAT TRANSFER

Problem 16

Determine the overall heat-transfer coefficient, U_o, based on the outer surface of a brass tube (2.5 cm inside diameter, 3.34 cm outside diameter, and thermal conductivity of 110 W/m·K) given the following conditions: the inside and outside heat-transfer coefficients are 1200 W/m²·K and 2000 W/m²·K, respectively, and the fouling factors for the inside and outside surfaces are 5500 W/m²·K each.

(A) 155 W/m²·K
(B) 480 W/m²·K
(C) 564 W/m²·K
(D) 642 W/m²·K

Solution

Convective heat-transfer, which occurs at the tube surfaces and is affected by fouling, is expressed in the Newton rate equation as

$$\frac{\dot{Q}}{A} = h\Delta T$$

The Fourier rate equation is used to calculate conductive heat-transfer in the radial direction through the tube wall.

$$\dot{Q}_r = -kA\Delta T$$

$$= -k2\pi rL\frac{dT}{dr}$$

$$\dot{Q}_r\int_{r_i}^{r_o}\frac{dr}{r} = -2\pi kL\int_{T_i}^{T_o}dT$$

$$\dot{Q}_r = \frac{2\pi kL}{\ln\dfrac{r_o}{r_i}}\Delta T$$

The overall heat-transfer coefficient, U, combines each individual thermal resistance present in a system. The Newton rate equation is expanded to calculate U using the outside area of the tube as a basis.

$$\dot{Q} = UA\Delta T = \frac{\Delta T}{\dfrac{1}{UA}} = \frac{\Delta T}{R_1 + R_2 + \cdots}$$

$$\frac{1}{U_o A_o} = R_1 + R_2 + \cdots$$

$$= \frac{1}{h_i A_i} + \frac{1}{h_{fi} A_i} + \frac{\ln\dfrac{D_o}{D_i}}{2\pi Lk} + \frac{1}{h_{fo} A_o} + \frac{1}{h_o A_o}$$

$$\frac{1}{U_o} = \frac{D_o}{h_i D_i} + \frac{D_o}{h_{fi} D_i} + \frac{D_o \ln \frac{D_o}{D_i}}{2k} + \frac{1}{h_{fo}} + \frac{1}{h_o}$$

$$= \frac{3.34 \text{ cm}}{\left(1200 \ \dfrac{\text{W}}{\text{m}^2 \cdot \text{K}}\right)(2.5 \text{ cm})}$$

$$+ \frac{3.34 \text{ cm}}{\left(5500 \ \dfrac{\text{W}}{\text{m}^2 \cdot \text{K}}\right)(2.5 \text{ cm})}$$

$$+ \frac{(0.0334 \text{ m}) \ln \left(\dfrac{3.34 \text{ cm}}{2.5 \text{ cm}}\right)}{(2)\left(110 \ \dfrac{\text{W}}{\text{m} \cdot \text{K}}\right)}$$

$$+ \frac{1}{5500 \ \dfrac{\text{W}}{\text{m}^2 \cdot \text{K}}} + \frac{1}{2000 \ \dfrac{\text{W}}{\text{m}^2 \cdot \text{K}}}$$

$$= 0.002\,08 \text{ m}^2 \cdot \text{K/W}$$

$$U_o = \left(\frac{1}{U_o}\right)^{-1} = \left(0.002\,08 \ \frac{\text{m}^2 \cdot \text{K}}{\text{W}}\right)^{-1}$$

$$= 480 \text{ W/m}^2 \cdot \text{K}$$

The answer is B.

Problems 17 and 18 are based on the following information.

After reaction, a liquid product (heat capacity = 1070 J/kg·K) is cooled from 370K to 325K in a counter-current, single-pass heat exchanger. The production rate is 2130 kg/h. Water (heat capacity = 4180 J/kg·K) is supplied as coolant at 0.185 kg/s and 275K. The overall heat-transfer coefficient is 242 W/m²·K.

Problem 17

The exit temperature of the water is most nearly

   (A) 290K
   (B) 310K
   (C) 320K
   (D) 340K

Solution

On the product (hot) side,

$$\dot{Q} = \dot{m} c_p \Delta T$$

$$= \left(\frac{2130 \ \dfrac{\text{kg}}{\text{h}}}{3600 \ \dfrac{\text{s}}{\text{h}}}\right)\left(1070 \ \frac{\text{J}}{\text{kg} \cdot \text{K}}\right)(370\text{K} - 325\text{K})$$

$$= 28\,489 \text{ W}$$

On the water (cold) side,

$$\dot{Q} = \dot{m} c_p (T_{Co} - T_{Ci})$$

$$T_{Co} = \frac{q}{\dot{m} c_p} + T_{Ci}$$

$$= \frac{28\,489 \text{ W}}{\left(0.185 \ \dfrac{\text{kg}}{\text{s}}\right)\left(4180 \ \dfrac{\text{J}}{\text{kg} \cdot \text{K}}\right)} + 275\text{K}$$

$$= 313\text{K} \quad (310\text{K})$$

The answer is B.

Problem 18

The heat-transfer surface area is most nearly

   (A) 1.81 m²
   (B) 1.89 m²
   (C) 2.18 m²
   (D) 2.86 m²

Solution

Heat flux for the heat exchanger is expressed as follows.

$$\dot{Q} = U A \Delta T_{\text{lm}}$$

$$= U A \left(\frac{(T_{Ho} - T_{Ci}) - (T_{Hi} - T_{Co})}{\ln\left(\dfrac{T_{Ho} - T_{Ci}}{T_{Hi} - T_{Co}}\right)}\right)$$

Rearranging,

$$A = \frac{\dot{Q}}{U}\left(\frac{\ln\left(\dfrac{T_{Ho} - T_{Ci}}{T_{Hi} - T_{Co}}\right)}{(T_{Ho} - T_{Ci}) - (T_{Hi} - T_{Co})}\right)$$

$$= \left(\frac{28\,500 \text{ W}}{242 \ \dfrac{\text{W}}{\text{m}^2 \cdot \text{K}}}\right)\left(\frac{\ln\left(\dfrac{325\text{K} - 275\text{K}}{370\text{K} - 312\text{K}}\right)}{\begin{array}{c}(325\text{K} - 275\text{K}) \\ - (370\text{K} - 312\text{K})\end{array}}\right)$$

$$= 2.18 \text{ m}^2$$

The answer is C.

Problem 19

A 1 cm thick aluminum plate-and-fin heat exchanger (thermal conductivity = 229 W/m·K) provides heat-transfer between water and air. The straight rectangular fins are 2.25 cm long and 0.125 cm thick, spaced 0.80 cm apart. The air-side heat-transfer coefficient and fin efficiency are 5.0 W/m²·K and 95%, respectively. On the water side they are 78.5 W/m²·K and 90%, respectively. Determine the increase in heat-transfer provided by placing these fins on the bare aluminum plate.

(A) 29.7%
(B) 92.5%
(C) 632%
(D) 662%

Solution

Heat transfer from a finned surface is

$$\dot{Q}_{\text{total}} = \dot{Q}_{\text{platesurface}} + \dot{Q}_{\text{fins}}$$
$$= A_o h (T_o - T_\infty) + A_F h (T - T_\infty)$$

The fin efficiency, η_F, is defined as the ratio of the actual heat transfer from the fin to the maximum possible heat transfer that would occur if the temperature along the fin remained constant, equal to the base temperature T_o. Rewriting the heat flux equation to incorporate fin efficiency yields

$$\dot{Q}_{\text{total}} = A_o h (T_o - T_\infty) + A_F h \eta_F (T_o - T_\infty)$$
$$= h (A_o + A_F \eta_F)(T_o - T_\infty)$$

Taking a 1 m × 1 m surface as a basis, the areas of the primary surface and fins are

$$A_o = \text{base area} - (\text{no. of fins}) \left(\frac{\text{surface area}}{\text{fins}} \right)$$

$$= 1 \text{ m}^2 - \left(\frac{1 \text{ m}}{0.008 \ \frac{\text{m}}{\text{fin}}} \right) \left((1 \text{ m}) \left(0.001\,25 \ \frac{\text{m}}{\text{fin}} \right) \right)$$

$$= 0.844 \text{ m}^2$$

$$A_F = (\text{no. of fins}) \left(\frac{\text{base area}}{\text{fins}} \right)$$

$$= (\text{no. of fins})(2LW + Wt)$$

$$= \left(\frac{1 \text{ m}}{0.008 \ \frac{\text{m}}{\text{fin}}} \right) \left(\begin{array}{l} (2)(1 \text{ m})(0.0225 \text{ m}) \\ + (1 \text{ m})(0.001\,25 \text{ m}) \end{array} \right)$$

$$= 5.78 \text{ m}^2$$

The air-side and water-side heat-transfer rates are

$$\dot{Q}_a = h_a (A_o + A_F \eta_{Fa})(T_o - T_a)$$
$$= \left(5.0 \ \frac{\text{W}}{\text{m}^2 \cdot \text{K}} \right) \left(\begin{array}{l} 0.844 \text{ m}^2 \\ + (5.78 \text{ m}^2)(0.95) \end{array} \right) (T_o - T_a)$$
$$= \left(31.7 \ \frac{\text{W}}{\text{K}} \right) (T_o - T_a) \quad [\text{per m}^2 \text{ base}]$$

$$\dot{Q}_w = h_w (A_o + A_F \eta_{Fw})(T_o - T_w)$$
$$= \left(78.5 \ \frac{\text{W}}{\text{m}^2 \cdot \text{K}} \right) \left(\begin{array}{l} 0.844 \text{ m}^2 \\ + (5.78 \text{ m}^2)(0.90) \end{array} \right) (T_o - T_w)$$
$$= \left(475 \ \frac{\text{W}}{\text{K}} \right) (T_o - T_w) \quad [\text{per m}^2 \text{ base}]$$

The thermal resistance per square meter in air and water is the reciprocal of the corresponding heat-transfer coefficient. The heat-transfer rate between the air and water is expressed as the overall temperature difference over the sum of thermal resistances.

$$\dot{Q}_{\text{finned}} = \frac{T_a - T_w}{\dfrac{1}{h_{a,F}} + \dfrac{L}{k} + \dfrac{1}{h_{w,F}}}$$

$$= \frac{T_a - T_w}{\dfrac{1}{31.7 \ \frac{\text{W}}{\text{m}^2 \cdot \text{K}}} + \dfrac{0.01 \text{ m}}{229 \ \frac{\text{W}}{\text{m}^2 \cdot \text{K}}} + \dfrac{1}{475 \ \frac{\text{W}}{\text{m}^2 \cdot \text{K}}}}$$

$$= \left(29.7 \ \frac{\text{W}}{\text{m}^2 \cdot \text{K}} \right) (T_a - T_w)$$

Without the fins, the heat flux is

$$\dot{Q}_{\text{nofins}} = \frac{T_a - T_w}{\dfrac{1}{h_a} + \dfrac{L}{k} + \dfrac{1}{h_w}}$$

$$= \frac{T_a - T_w}{\dfrac{1}{5.0 \ \frac{\text{W}}{\text{m}^2 \cdot \text{K}}} + \dfrac{0.01 \text{ m}}{229 \ \frac{\text{W}}{\text{m}^2 \cdot \text{K}}} + \dfrac{1}{78.5 \ \frac{\text{W}}{\text{m}^2 \cdot \text{K}}}}$$

$$= \left(4.70 \ \frac{\text{W}}{\text{m}^2 \cdot \text{K}} \right) (T_a - T_w)$$

The increased heat flux is

$$\frac{\dot{Q}_{\text{finned}}}{\dot{Q}_{\text{nofins}}} = \frac{\left(29.7 \ \frac{\text{W}}{\text{m}^2 \cdot \text{K}} \right) (T_a - T_w)}{\left(4.70 \ \frac{\text{W}}{\text{m}^2 \cdot \text{K}} \right) (T_a - T_w)}$$

$$= 6.32 \quad (632\%)$$

The answer is C.

Problems 20 and 21 are based on the following information and illustration.

A food company produces spaghetti with meatballs. The sauce, pasta, and meatball components are cooked separately and then mixed before canning. Meatballs are rolled to a uniform 2.0 cm diameter and held at a constant 8°C prior to cooking. They are steam-cooked in an environment where water vapor at 100°C condenses on the meatball surface with an effective film coefficient, h, of 8500 W/m²·K. Assume the meatballs are uniform with the following constant thermal characteristics.

$$\text{thermal conductivity} = 0.45 \text{ W/m·K}$$
$$\text{density} = 750 \text{ kg/m}^3$$
$$\text{heat capacity} = 3.0 \text{ kJ/kg·K}$$

The following logarithmic temperature history chart for the center of a meatball may be used when internal temperature resistance dominates.

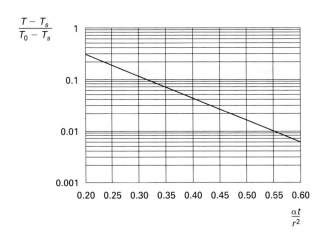

Problem 20

In order to neglect internal temperature resistance involved in heating a body, the Biot number must be

(A) less than 0.1
(B) less than 1
(C) greater than 1
(D) Bi is not a factor.

Solution

The Biot number is a ratio of the conductive (internal) thermal resistance to the convective (external) thermal resistance.

$$\text{Bi} = \frac{h\dfrac{V}{A}}{k}$$

When the Biot number is much less than 1.0 (typically 0.1 or less), surface resistance dominates and internal resistance can be neglected.

The answer is A.

Problem 21

In order to prevent botulism, the center of a meatball must reach 90°C. How much time is required for the meatballs to cook?

(A) 150 s
(B) 212 s
(C) 250 s
(D) 298 s

Solution

The temperature history chart must be used if internal temperature resistance dominates; that is, if Bi > 0.1.

$$\text{Bi} = \frac{h\dfrac{V}{A}}{k}$$

$$= \frac{\left(8500 \ \dfrac{\text{W}}{\text{m}^2\cdot\text{K}}\right)\left(\dfrac{4}{3}\right)\pi(0.01 \ \text{m})^3}{\left(0.45 \ \dfrac{\text{W}}{\text{m}^2\cdot\text{K}}\right)4\pi(0.01 \ \text{m})^2}$$

$$= 63.0 > 0.1$$

$$\frac{T - T_s}{T_0 - T_s} = \frac{90°\text{C} - 100°\text{C}}{8°\text{C} - 100°\text{C}}$$

$$= 0.11$$

The corresponding abscissa value is approximately 0.30. The time can be calculated.

$$0.30 = \frac{\alpha t}{r^2} = \left(\frac{k}{\rho c_p}\right)\left(\frac{t}{r^2}\right)$$

$$t = \frac{0.30 r^2 \rho c_p}{k}$$

$$= \frac{(0.30)(0.01 \ \text{m})^2 \left(750 \ \dfrac{\text{kg}}{\text{m}^3}\right)\left(3000 \ \dfrac{\text{J}}{\text{kg}\cdot\text{K}}\right)}{0.45 \ \dfrac{\text{W}}{\text{m}^2\cdot\text{K}}}$$

$$= 150 \ \text{s}$$

The answer is A.

MASS TRANSFER

Problems 22–25 are based on the following information.

Air flows through a combustion chamber held at 1145K and 101.3 kPa. Oxygen diffuses through a 1.5 mm thick stagnant film of air to a flat carbon surface where it reacts instantaneously to form carbon dioxide. No reaction occurs in the gas film. The diffusivity of oxygen in the gas film is $1.30 \times 10^{-4} \ \text{m}^2/\text{s}$.

Problem 22

The rate of oxygen diffusion to the surface is most nearly

(A) 0.176 mol/m²·s
(B) 0.194 mol/m²·s
(C) 0.208 mol/m²·s
(D) 0.922 mol/m²·s

Solution

The rate of diffusion is expressed as the molar flux of a component in the z-direction.

$$N_A = -CD_m \frac{dy_A}{dz} + y_A \sum_{i=1}^{n} N_i$$

For diffusion of two species expressed in terms of partial pressure,

$$N_A = \left(\frac{D_m}{RT}\right)\left(\frac{dp_A}{dz}\right) + \frac{p_A}{p}(N_A + N_B)$$

The reaction at the surface is

$$O_2 + C \longrightarrow CO_2$$

One mole of O_2 forms one mole of CO_2. Therefore,

$$N_{O_2} = -N_{CO_2}$$
$$= \left(\frac{D_m}{RT}\right)\left(\frac{dp_{O_2}}{dz}\right) + \frac{p_{O_2}}{p}(N_{O_2} - N_{O_2})$$
$$= \left(\frac{D_m}{RT}\right)\left(\frac{dp_{O_2}}{dz}\right)$$

Separating variables and integrating yields

$$N_{O_2}\int_{z_1}^{z_2} dz = \frac{D_m}{RT}\int_{p_{1,O_2}}^{p_{2,O_2}} dp_{O_2}$$
$$N_{O_2}(z_2 - z_1) = \frac{D_m}{RT}(p_{2,O_2} - p_{1,O_2})$$

The partial pressure of oxygen at the carbon surface (p_{1,O_2}) equals 0 because the reaction is instantaneous. The oxygen partial pressure at the top of the stagnant gas film equals the partial pressure of oxygen in air.

$$p_{2,O_2} = (0.21)(101.3 \text{ kPa}) = 21.3 \text{ kPa}$$
$$N_{O_2} = \left(\frac{D_m}{RT}\right)\left(\frac{p_{2,O_2} - p_{1,O_2}}{z_2 - z_1}\right)$$
$$= \left(\frac{1.30 \times 10^{-4} \frac{\text{m}^2}{\text{s}}}{\left(8.314 \frac{\text{m}^3 \cdot \text{Pa}}{\text{mol} \cdot \text{K}}\right)(1145\text{K})}\right)$$
$$\times \left(\frac{21\,300 \text{ Pa} - 0}{1.5 \times 10^{-3} \text{ m} - 0}\right)$$
$$= 0.194 \text{ mol/m}^2 \cdot \text{s}$$

The answer is B.

Problem 23

The partial pressure of carbon dioxide 1 mm from the carbon surface is most nearly

(A) 7.1 kPa
(B) 14 kPa
(C) 21 kPa
(D) 34 kPa

Solution

The concentration profile through the film is calculated after considering the general differential equation for mass transfer.

$$\nabla N_A + \frac{\partial C_A}{\partial t} - R_A = 0$$

The second term can be eliminated at steady state, and the third term can be eliminated because the reaction occurs as a boundary condition and not within the gas film.

$$\nabla N_A = \frac{dN_A}{dz} = 0$$

From Prob. 22,

$$N_A = \left(\frac{D_m}{RT}\right)\left(\frac{dp_A}{dz}\right)$$

Therefore,

$$\frac{dN_A}{dz} = \left(\frac{D_m}{RT}\right)\left(\frac{d^2 p_A}{dz^2}\right) = 0$$
$$\frac{d^2 p_A}{dz^2} = 0$$
$$p_A = C_1 z + C_2$$

The boundary conditions for CO_2 are as follows.

At $z = 0$ mm:

$$p_{CO_2} = 21.3 \text{ kPa}$$

At $z = 1.5$ mm:

$$p_{CO_2} = 0 \text{ kPa}$$

Therefore,

$$p_{CO_2} = 21.3 \text{ kPa} - \left(14.2 \frac{\text{kPa}}{\text{mm}}\right)z$$

At $z = 1$ mm:

$$p_{CO_2} = 21.3 \text{ kPa} - \left(14.2 \frac{\text{kPa}}{\text{mm}}\right)(1 \text{ mm})$$
$$= 7.1 \text{ kPa}$$

The answer is A.

Problem 24

The mass-transfer coefficient for oxygen in the gas film is most nearly

(A) 1.30×10^{-4} m^2/s
(B) 1.38×10^{-3} mol/m·s
(C) 8.67×10^{-2} m/s
(D) 0.922 mol/m^2·s

Solution

In an analogous manner to heat and momentum transfer, a mass-transfer coefficient relates molar flux to a concentration gradient.

$$N_A = k_m(C_{A_2} - C_{A_1})$$
$$= \frac{k_m}{RT}(p_{A_2} - p_{A_1})$$
$$= \left(\frac{D_m}{RT}\right)\left(\frac{p_{A_2} - p_{A_1}}{z_2 - z_1}\right)$$

Rearranging,

$$k_m = \frac{D_m}{z_2 - z_1}$$
$$= \frac{1.3 \times 10^{-4} \frac{m^2}{s}}{1.5 \times 10^{-3} \text{ m} - 0 \text{ m}}$$
$$= 8.7 \times 10^{-2} \text{ m/s}$$

The answer is C.

Problem 25

If the oxygen reacts instantaneously with the carbon surface to form carbon monoxide instead of carbon dioxide (all other parameters remaining unchanged), the rate of oxygen diffusion to the surface is most nearly

(A) $0.176 \text{ mol/m}^2\text{·s}$
(B) $0.194 \text{ mol/m}^2\text{·s}$
(C) $0.208 \text{ mol/m}^2\text{·s}$
(D) $0.922 \text{ mol/m}^2\text{·s}$

Solution

When oxygen reacts to form carbon monoxide, two moles of CO diffuse from the surface for each mole of O_2 diffusing to the surface.

$$O_2 + 2C \longrightarrow 2CO$$
$$2N_{O_2} = -N_{CO}$$
$$N_{O_2} = \left(\frac{D_m}{RT}\right)\left(\frac{dp_{O_2}}{dz}\right) + \left(\frac{p_{O_2}}{p}\right)$$
$$\times (N_{O_2} - 2N_{O_2})$$
$$N_{O_2}\left(1 + \frac{p_{O_2}}{p}\right) = \left(\frac{D_m}{RT}\right)\left(\frac{dp_{O_2}}{dz}\right)$$

Separating variables and integrating,

$$N_{O_2}\int_{z_1}^{z_2} dz = \frac{D_m}{RT}\int_{p_{O_2,1}}^{p_{O_2,z2}} \frac{dp_{O_2}}{1 + \frac{p_{O_2}}{p}}$$
$$N_{O_2}(z_2 - z_1) = \left(\frac{pD_m}{RT}\right)\ln\left(\frac{p + p_{O_2,2}}{p + p_{O_2,1}}\right)$$

$$N_{O_2} = \left(\frac{pD_m}{RT(z_2 - z_1)}\right)\ln\left(\frac{p + p_{O_2,2}}{p + p_{O_2,1}}\right)$$
$$= \left(\frac{(1.013 \times 10^5 \text{ Pa})\left(1.30 \times 10^{-4} \frac{m^2}{s}\right)}{\left(8.314 \frac{m^3\cdot Pa}{mol\cdot K}\right)(1145K)(1.5 \times 10^{-3} \text{ m} - 0)}\right)$$
$$\times \ln\left(\frac{101.3 \text{ kPa} + 21.3 \text{ kPa}}{101.3 \text{ kPa} + 0}\right)$$
$$= 0.176 \text{ mol/m}$$

The answer is A.

Problems 26 and 27 are based on the following information.

A packed column is used to reduce the levels of a pollutant in a gas stream from a mole fraction of 0.025 to a mole fraction of 0.000 15. The gas stream flows at a rate of 10.0 m^3/min while pure water enters the top of the column at a rate of 15.0 kg/min. The 2.25 m diameter tower will be packed with 19 mm ceramic Raschig rings, providing an interfacial surface area-to-volume ratio of 262 m^2/m^3. The overall mass-transfer coefficient based on the gas-phase driving force is 69.4 mol/m^2·h. The system operates at 40°C and atmospheric pressure. The pollutant follows the Henry's law relation with the Henry's law constant equal to 1.75×10^5 Pa.

Problem 26

The pollutant mole fraction in the exiting water stream is most nearly

(A) 8.68×10^{-5}
(B) 0.0116
(C) 0.0145
(D) 0.0533

Solution

The composition of the exiting liquid stream is determined by mass balance. The liquid and gas stream flow rates can be assumed constant throughout the column.

$$Lx_{in} + Gy_{in} = Lx_{out} + Gy_{out}$$

The mole fraction of pollutant entering in the water stream is 0 (that is, $x_{in} = 0$), so

$$x_{out} = \frac{G}{L}(y_{in} - y_{out})$$

The gas and liquid molar velocities are

$$G = \left(\frac{pV}{RT}\right)\left(\frac{1}{A}\right)$$

$$= \frac{(1.013 \times 10^5 \text{ Pa})\left(10.0 \frac{\text{m}^3}{\text{min}}\right)\left(60 \frac{\text{min}}{\text{h}}\right)}{\left(8.314 \frac{\text{m}^3 \cdot \text{Pa}}{\text{mol} \cdot \text{K}}\right)(313\text{K})\left(\frac{\pi(2.25 \text{ m})^2}{4}\right)}$$

$$= 5874 \text{ mol/m}^2 \cdot \text{h}$$

$$L = \frac{\left(15.0 \frac{\text{kg}}{\text{min}}\right)\left(60 \frac{\text{min}}{\text{h}}\right)}{\left(0.0180 \frac{\text{kg}}{\text{mol}}\right)\left(\frac{\pi(2.25 \text{ m})^2}{4}\right)}$$

$$= 12\,575 \text{ mol/m}^2 \cdot \text{h}$$

$$x_{\text{out}} = \frac{G}{L}(y_{\text{in}} - y_{\text{out}})$$

$$= \left(\frac{5874 \frac{\text{mol}}{\text{m}^2 \cdot \text{h}}}{12\,575 \frac{\text{mol}}{\text{m}^2 \cdot \text{h}}}\right)(0.025 - 0.00015)$$

$$= 0.0116$$

The answer is B.

Problem 27

The total height of packing required to achieve the desired separation is most nearly

(A) 1.61 m
(B) 5.81 m
(C) 20.3 m
(D) 53.5 m

Solution

The total height of packing required is the product of the transfer unit height and the number of transfer units.

$$z = (\text{HTU})(\text{NTU})$$

$$\text{NTU} = \frac{y_1 - y_2}{(y - y^*)_{\text{lm}}}$$

The log-mean driving force is

$$(y - y^*)_{\text{lm}} = \frac{(y_1 - y_1^*) - (y_2 - y_2^*)}{\ln\left(\frac{y_1 - y_1^*}{y_2 - y_2^*}\right)}$$

The equilibrium concentration of pollutant in the gas stream as it exits the column (subscript 2) is 0 as the entering water is pollutant free. As the gas stream enters the column (subscript 1), the equilibrium concentration of pollutant is calculated using Henry's law.

$$y_1^* = \frac{H}{p}x_1$$

$$= \left(\frac{1.75 \times 10^5 \text{ Pa}}{1.013 \times 10^5 \text{ Pa}}\right)(0.0116)$$

$$= 0.0200$$

The log-mean driving force is

$$(y - y^*)_{\text{lm}} = \frac{(0.025 - 0.0200) - (0.000\,15 - 0)}{\ln\left(\frac{0.025 - 0.0200}{0.000\,15 - 0}\right)}$$

$$= 1.38 \times 10^{-3}$$

Calculate the transfer unit height. The height of a transfer function unit is the ratio of the gas velocity to the overall gas capacity coefficient. The coefficient is the product of the overall mass-transfer coefficient and the interfacial surface area-to-volume ratio.

$$\text{HTU} = \frac{G}{K_G A}$$

$$= \frac{5874 \frac{\text{mol}}{\text{m}^2 \cdot \text{h}}}{\left(69.4 \frac{\text{mol}}{\text{m}^2 \cdot \text{h}}\right)\left(262 \frac{\text{m}^2}{\text{m}^3}\right)}$$

$$= 0.323 \text{ m}$$

Next, calculate the number of transfer units.

$$\text{NTU} = \frac{y_1 - y_2}{(y - y^*)_{\text{lm}}}$$

$$= \frac{0.025 - 0.00015}{1.38 \times 10^{-3}}$$

$$= 18.0$$

The total height of packing required is

$$z = (\text{HTU})(\text{NTU})$$

$$= (0.323 \text{ m})(18.0)$$

$$= 5.81 \text{ m}$$

The answer is B.

MATERIAL AND ENERGY BALANCES

Problems 28–31 are based on the following information.

Aspirin is made in an aqueous reaction between salicylic acid and acetic anhydride according to the following reaction.

$$2C_7H_6O_3(aq) + C_4H_6O_3(l)$$
$$\longrightarrow 2C_9H_8O_4(aq) + H_2O(l)$$

Feed streams of 27.2 kg/h salicylic acid in 74.7 kg/h water and 15.1 kg/h acetic anhydride join with a recycle stream before entering the reactor. Reaction proceeds to 50% conversion of the acetic anhydride. The stream exiting the reactor passes through a separator that yields a product stream containing 34.9 kg/h aspirin and the recycle stream, which contains no aspirin. The recycle stream contains equal moles of salicylic acid and acetic anhydride, and it circulates water at 18.0 kg/h.

Each of the stream flow rates and compositions can be determined from a series of mass balances. The following flowchart illustrates the given information.

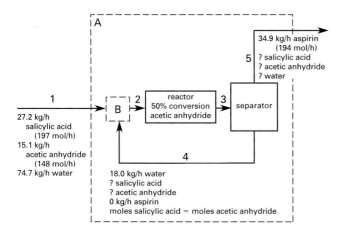

The molar flow rates of reactants are calculated from mass flow rates and compound molecular weights as follows.

Feed:

$$\text{mol salicylic acid} = \frac{27\,200\ \dfrac{g}{h}}{138\ \dfrac{g}{mol}}$$

$$= 197\ \text{mol/h}$$

$$\text{mol acetic acid} = \frac{15\,100\ \dfrac{g}{h}}{102\ \dfrac{g}{mol}}$$

$$= 148\ \text{mol/h}$$

$$\text{mol water} = \frac{74\,700\ \dfrac{g}{h}}{18.0\ \dfrac{g}{mol}}$$

$$= 4150\ \text{mol/h}$$

Product:

$$\text{mol aspirin} = \frac{34\,900\ \dfrac{g}{h}}{180\ \dfrac{g}{mol}}$$

$$= 194\ \text{mol/h}$$

The overall component mass balances around process A (see illustration) determine the exit compositions.

$$\text{moles in} + \text{moles generated}$$
$$= \text{moles reacted} + \text{moles out}$$

$$\text{moles out} = \text{moles in} + \text{moles generated}$$
$$- \text{moles reacted}$$

Salicylic acid:

$$\text{moles out} = 197\ \frac{\text{mol}}{h} + 0\ \frac{\text{mol}}{h} - 194\ \frac{\text{mol}}{h}$$

$$= 3\ \text{mol/h}$$

Acetic anhydride:

$$\text{moles out} = 148\ \frac{\text{mol}}{h} + 0\ \frac{\text{mol}}{h} - \frac{194\ \dfrac{\text{mol}}{h}}{2}$$

$$= 51\ \text{mol/h}$$

Water:

$$\text{moles out} = 4150\ \frac{\text{mol}}{h} + \frac{194\ \dfrac{\text{mol}}{h}}{2} - 0\ \frac{\text{mol}}{h}$$

$$= 4247\ \text{mol/h}$$

Using acetic anhydride as the tie component, calculate the compositions of the streams.

- Aspirin component balance around separator:

$$\text{mol}_3 = \text{mol}_4 + \text{mol}_5$$

$$= 0\ \frac{\text{mol}}{h} + 194\ \frac{\text{mol}}{h}$$

$$= 194\ \text{mol/h}$$

- 50% conversion of acetic anhydride:

$$(0.5)(\text{mol}_{2(\text{ac.an.})}) = \text{moles reacted}$$

$$= 0.5\ \text{mol}_{3(\text{asp.})}$$

$$\text{mol}_{2(\text{ac.an.})} = \text{mol}_{3(\text{asp.})}$$

$$= 194\ \text{mol/h}$$

$$\text{mol}_{3(\text{ac.an.})} = (0.5)(\text{mol}_{2(\text{ac.an.})})$$

$$= 97\ \text{mol/h}$$

- Acetic anhydride component balance around separator:

$$\text{mol}_4 = \text{mol}_3 - \text{mol}_5$$

$$= 97\ \frac{\text{mol}}{h} - 51\ \frac{\text{mol}}{h}$$

$$= 46\ \text{mol/h}$$

- The problem states that the number of moles of acetic anhydride and salicylic acid in the recycle stream are equal. Therefore,

$$\text{mol}_{4(\text{sa.ac.})} = \text{mol}_{4(\text{ac.an.})} = 46 \text{ mol/h}$$

Component balances around the separator define stream 3.

Salicylic acid:

$$\text{mol}_3 = \text{mol}_4 + \text{mol}_5$$
$$= 46 \frac{\text{mol}}{\text{h}} + 3 \frac{\text{mol}}{\text{h}}$$
$$= 49 \text{ mol/h}$$

Water:

$$\text{mol}_3 = \text{mol}_4 + \text{mol}_5$$
$$= \frac{18\,000 \frac{\text{g}}{\text{h}}}{18.0 \frac{\text{g}}{\text{mol}}} + 4247 \frac{\text{mol}}{\text{h}}$$
$$= 5247 \text{ mol/h}$$

Component balances around the mixing point B (see illustration) define stream 2.

Salicylic acid:

$$\text{mol}_2 = \text{mol}_1 + \text{mol}_4$$
$$= 197 \frac{\text{mol}}{\text{h}} + 46 \frac{\text{mol}}{\text{h}}$$
$$= 243 \text{ mol/h}$$

Water:

$$\text{mol}_2 = \text{mol}_1 + \text{mol}_4$$
$$= 4150 \frac{\text{mol}}{\text{h}} + \frac{18\,000 \frac{\text{g}}{\text{h}}}{18.0 \frac{\text{g}}{\text{mol}}}$$
$$= 5150 \text{ mol/h}$$

Problem 28

The flow rate from the reactor to the separator is most nearly

(A) 117 kg/h
(B) 146 kg/h
(C) 223 kg/h
(D) 340 kg/h

Solution

The mass flow rate from the reactor to the separator (stream 3) is calculated by converting the molar flow rates to mass flow rates and summing over the components.

$$Q_3 = \left(\left(49 \frac{\text{mol}_{\text{sa.ac.}}}{\text{h}} \right) \left(138 \frac{\text{g}}{\text{mol}} \right) \right.$$
$$+ \left(97 \frac{\text{mol}_{\text{ac.an.}}}{\text{h}} \right) \left(102 \frac{\text{g}}{\text{mol}} \right)$$
$$+ \left(5247 \frac{\text{mol}_{\text{water}}}{\text{h}} \right) \left(18 \frac{\text{g}}{\text{mol}} \right)$$
$$+ \left. \left(194 \frac{\text{mol}_{\text{asp.}}}{\text{h}} \right) \left(180 \frac{\text{g}}{\text{mol}} \right) \right) \left(\frac{1 \text{ kg}}{1000 \text{ g}} \right)$$
$$= 146 \text{ kg/h}$$

The answer is B.

Problem 29

The recycle ratio is most nearly

(A) 0.154
(B) 0.199
(C) 0.248
(D) 0.371

Solution

The recycle ratio is the mass ratio of recycle to outlet.

$$R = \frac{Q_4}{Q_5}$$
$$Q_5 = Q_1$$
$$= 117 \text{ kg/h}$$
$$Q_4 = Q_3 - Q_5$$
$$= 146 \frac{\text{kg}}{\text{h}} - 117 \frac{\text{kg}}{\text{h}}$$
$$= 29 \text{ kg/h}$$
$$R = \frac{Q_4}{Q_5}$$
$$= \frac{29 \text{ kg}}{117 \text{ kg}}$$
$$= 0.248$$

The answer is C.

Problem 30

The composition of the recycle stream is most nearly

(A) 18.1 wt% salicylic acid, 18.1 wt% acetic anhydride, and 63.8 wt% water
(B) 21.9 wt% salicylic acid, 16.2 wt% acetic anhydride, and 62.1 wt% water
(C) 23.2 wt% salicylic acid, 12.9 wt% acetic anhydride, and 63.8 wt% water
(D) 53.5 wt% salicylic acid, 39.5 wt% acetic anhydride, and 6.98 wt% water

Solution

The composition of the recycle stream is

$$\text{wt\%}_{\text{sa.ac.}} = \frac{\text{wt}_{\text{sa.ac.}}}{Q_4} \times 100\%$$

$$= \left(\frac{\left(46 \ \frac{\text{mol}}{\text{h}}\right) \left(138 \ \frac{\text{g}}{\text{mol}}\right)}{29\,000 \ \frac{\text{g}}{\text{h}}} \right) \times 100\%$$

$$= 21.9 \ \text{wt\%}$$

$$\text{wt\%}_{\text{ac.an.}} = \frac{\text{wt}_{\text{ac.an.}}}{Q_4} \times 100\%$$

$$= \frac{\left(46 \ \frac{\text{mol}}{\text{h}}\right) \left(102 \ \frac{\text{g}}{\text{mol}}\right)}{29\,000 \ \frac{\text{g}}{\text{h}}} \times 100\%$$

$$= 16.2 \ \text{wt\%}$$

$$\text{wt\%}_{\text{water}} = \frac{\text{wt}_{\text{water}}}{Q_4} \times 100\%$$

$$= \left(\frac{18 \ \frac{\text{kg}}{\text{h}}}{29 \ \frac{\text{kg}}{\text{h}}} \right) \times 100\%$$

$$= 62.1 \ \text{wt\%}$$

The answer is B.

Problem 31

The conversion of salicylic acid in the reactor is most nearly

  (A) 20.2%
  (B) 50.0%
  (C) 79.8%
  (D) 98.5%

Solution

The conversion of salicylic acid is the ratio of moles reacted to moles charged.

$$\left(\frac{\text{mol}_2 - \text{mol}_3}{\text{mol}_2} \right) \times 100\%$$

$$= \left(\frac{243 \ \frac{\text{mol}}{\text{h}} - 49 \ \frac{\text{mol}}{\text{h}}}{243 \ \frac{\text{mol}}{\text{h}}} \right) \times 100\%$$

$$= 79.8\%$$

The answer is C.

Problems 32 and 33 are based on the following information.

A tank contains 23.7 wt% sodium hydroxide in water. After adding 45.1 kg of NaOH pellets, the solution is 34.2 wt% caustic.

Problem 32

The original mass of solution is most nearly

  (A) 102 kg
  (B) 283 kg
  (C) 350 kg
  (D) 385 kg

Solution

Designating x (in kg) as the initial mass in the tank, $x + 45.1$ kg is the final weight. A NaOH component balance yields

$$\text{original mass} + \text{added mass} = \text{final mass}$$

$$0.237x + 45.1 \ \text{kg} = 0.342(x + 45.1 \ \text{kg})$$

$$x = \frac{(45.1 \ \text{kg})(1 - 0.342)}{0.342 - 0.237}$$

$$= 283 \ \text{kg}$$

The answer is B.

Problem 33

The heat of solution for sodium hydroxide in water is -42.6 kJ/mol. The heat change after adding the pellets is most nearly

  (A) 48 MJ generated
  (B) 48 MJ absorbed
  (C) 120 MJ generated
  (D) 120 MJ absorbed

Solution

Change in heat is calculated from the heat of solution. A negative sign indicates that heat is generated. The molecular weight of NaOH is 40 g/mol.

$$Q = \left(-42.6 \ \frac{\text{kJ}}{\text{mol}} \right) (45.1 \ \text{kg}) \left(\frac{1 \ \text{mol}}{0.0400 \ \text{kg}} \right)$$

$$= -48\,032 \ \text{kJ} \quad (48 \ \text{MJ}) \quad [\text{generated}]$$

The answer is A.

Problem 34

0.250 m^3 of water at $20°C$ is heated to $100°C$ by injecting saturated steam at $100°C$. Calculate the amount of steam needed if heat losses total 3750 kJ.

(A) 1.67 kg
(B) 8.86 kg
(C) 10.5 kg
(D) 38.7 kg

Solution

Steam requirements are calculated with an energy balance. The enthalpy change from condensing the steam will equal the heat losses plus the enthalpy change from heating the water.

$$m_{steam}h_{fg} = \text{losses} + m_w c_p \Delta T$$

$$m_{steam} = \frac{\text{losses} + m_w c_p \Delta T}{h_{fg}}$$

$$= \frac{3750 \text{ kJ} + (0.250 \text{ m}^3)\left(1000 \frac{\text{kg}}{\text{m}^3}\right)}{2257 \frac{\text{kJ}}{\text{kg}}}$$
$$\times \left(4.18 \frac{\text{kJ}}{\text{kg·K}}\right)(100°\text{C} - 20°\text{C})$$

$$= 38.7 \text{ kg}$$

The answer is D.

Problems 35 and 36 are based on the following information.

Two evaporators, A and B, are used to concentrate a 50% aqueous KCl solution to a 90% aqueous KCl solution. The feed stream 1 has a flow rate of 1000 kg/h.

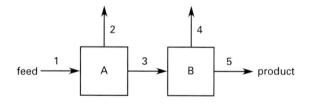

Problem 35

If the flow rate of stream 2 is twice the flow rate of stream 4, what is most nearly the flow rate of stream 4?

(A) 135 kg/h
(B) 146 kg/h
(C) 148 kg/h
(D) 153 kg/h

Solution

Stream 1 is 50% KCl and 50% water. At 1000 kg total, the KCl and water flow rates are 500 kg/h each.

Streams 3 and 5 also have a 500 kg/h KCl flow rate, since KCl is nonvolatile.

Stream 5 contains 90% KCl. The total flow rate of stream 5 is

$$Q_5 = \frac{Q_{KCl\ 5}}{0.90} = \frac{500 \frac{\text{kg}}{\text{h}}}{0.90}$$
$$= 555.5 \text{ kg/h}$$

The amount of water flowing out of stream 5 is the total flow rate of stream 5 minus the amount of KCl.

$$Q_{H_2O,\text{out } 5} = Q_5 - Q_{KCl\ 5}$$
$$= 555.5 \frac{\text{kg}}{\text{h}} - 500 \frac{\text{kg}}{\text{h}}$$
$$= 55.5 \text{ kg/h}$$

The amount of water flowing in streams 2 and 4 is the amount of water input, 500 kg/h, less the amount of water output, 55.5 kg/h.

$$Q_{H_2O,2\ 4} = Q_{H_2O,\text{in } 1\ 5} - Q_{H_2O,\text{out } 2\ 4}$$
$$= 500 \frac{\text{kg}}{\text{h}} - 55.5 \frac{\text{kg}}{\text{h}}$$
$$= 444.5 \text{ kg/h}$$

Since the flow rate of stream 2 is twice the flow rate of stream 4, let Q_4 equal the flow rate of stream 4 (100% water). Q_4 is found by

$$444.5 \frac{\text{kg}}{\text{h}} = 2Q_4 + Q_4$$
$$= 3Q_4$$

Rearranging to solve for Q_4,

$$Q_4 = \frac{444.5 \frac{\text{kg}}{\text{h}}}{3}$$
$$= 148.2 \text{ kg/h} \quad (148 \text{ kg/h})$$

The answer is C.

Problem 36

What is most nearly the concentration of KCl in stream 3 if the flow rate of stream 4 is 200 kg/h?

(A) 66%
(B) 71%
(C) 76%
(D) 80%

Solution

$$Q_1 = 1000 \text{ kg/h}$$

Stream 1 is 50% KCl and 50% H_2O; therefore the flow rates of KCl and H_2O are

$$Q_{\text{KCl } 1} = \left(1000 \ \frac{\text{kg}}{\text{h}}\right)(0.50)$$
$$= 500 \text{ kg/h}$$
$$Q_{H_2O \ 1} = 500 \text{ kg/h}$$

Stream 5 contains 500 KCl since KCl is nonvolatile. Stream 5 has a concentration of 90% KCl. Therefore,

$$Q_3 = \frac{Q_{\text{KCl } 5}}{0.90} = \frac{500 \ \frac{\text{kg}}{\text{h}}}{0.90}$$
$$= 555.5 \text{ kg/h}$$

Given that stream 4 is 200 kg/h, stream 3 can be found by the following equation.

$$Q_3 = Q_4 + Q_5$$
$$= 200 \ \frac{\text{kg}}{\text{h}} + 555.5 \ \frac{\text{kg}}{\text{h}}$$
$$= 755.5 \text{ kg/h}$$

Stream 3 contains 500 KCl. Therefore, the concentration of KCl in stream 3 is

$$C_{\text{KCl } 3} = \frac{Q_{\text{KCl } 3}}{Q_3}$$
$$= \frac{500 \ \frac{\text{kg}}{\text{h}}}{755.5 \ \frac{\text{kg}}{\text{h}}} \times 100\%$$
$$= 66.2\% \quad (66\%)$$

The answer is A.

SAFETY, HEALTH, AND ENVIRONMENTAL

Problems 37–38 are based on the following information.

The off-gas from a petroleum refinery is burned in a flare at 25°C and 101.3 kPa with 30% excess air to ensure complete combustion. The off-gas is composed at an elemental level of 87.2 wt% carbon, 12.6 wt% hydrogen, and 0.22 wt% sulfur. The exit gas from the flare is passed through a scrubber where water flowing counter-currently absorbs 80% of the sulfur dioxide. The water exiting the scrubber contains 0.375 wt% sulfur dioxide.

Problem 37

How much air is supplied to the flare?

   (A)   3.30 $\text{m}^3/\text{kg}_{\text{off-gas}}$
   (B)   12.1 $\text{m}^3/\text{kg}_{\text{off-gas}}$
   (C)   15.8 $\text{m}^3/\text{kg}_{\text{off-gas}}$
   (D)   20.5 $\text{m}^3/\text{kg}_{\text{off-gas}}$

Solution

The combustion reactions are

$$C + O_2 \longrightarrow CO_2$$
$$2H_2 + O_2 \longrightarrow 2H_2O$$
$$S + O_2 \longrightarrow SO_2$$

Convert the composition data to moles. Calculate the stoichiometric oxygen requirement as tabulated.

compound	wt%	amount per kilogram of gas (mol)	O_2 required (mol)
C (as C)	87.2	$\left(872 \ \frac{\text{g}}{\text{kg}_{\text{gas}}}\right)\left(\frac{1 \text{ mol}}{12.0 \text{ g}}\right) = 72.7$	72.7
H (as H_2)	12.6	$\left(126 \ \frac{\text{g}}{\text{kg}_{\text{gas}}}\right)\left(\frac{1 \text{ mol}}{2.02 \text{ g}}\right) = 62.4$	$\frac{62.4}{2} = 31.2$
S (as S)	0.22	$\left(2.2 \ \frac{\text{g}}{\text{kg}_{\text{gas}}}\right)\left(\frac{1 \text{ mol}}{32.1 \text{ g}}\right) = 0.0685$	0.0685
total			104

The stoichiometric requirement for oxygen is 104 mol. Air is 21 mol% oxygen. Including the 30% excess, the air requirement is

$$\text{air} = (1.3)\left(104 \ \frac{\text{mol}_{O_2}}{\text{kg}_{\text{off-gas}}}\right)\left(\frac{1 \text{ mol}_{\text{air}}}{0.21 \text{ mol}_{O_2}}\right)$$
$$= 644 \text{ mol}_{\text{air}}/\text{kg}_{\text{off-gas}}$$

The volume is calculated from the ideal gas law.

$$pV = nRT$$
$$V = \frac{nRT}{p}$$
$$= \frac{\left(644 \ \frac{\text{mol}_{\text{air}}}{\text{mol}_{\text{off-gas}}}\right)\left(8.314 \ \frac{\text{m}^3 \cdot \text{Pa}}{\text{mol} \cdot \text{K}}\right) \times (25°C + 273°)}{1.013 \times 10^5 \text{ Pa}}$$
$$= 15.8 \text{ m}^3/\text{kg}_{\text{off-gas}}$$

The answer is C.

Problem 38

The mole fraction of sulfur dioxide in the gas exiting the flare is most nearly

(A) 8.80×10^{-5}
(B) 1.01×10^{-4}
(C) 8.05×10^{-4}
(D) 2.20×10^{-3}

Solution

The total number of moles exiting the flare must be determined to calculate a mole fraction. Sum the components.

$$CO_2 = \text{mol C entering}$$
$$= 72.7 \text{ mol}$$
$$H_2O = \text{mol } H_2 \text{ entering}$$
$$= 62.4 \text{ mol}$$
$$SO_2 = \text{mol S entering}$$
$$= 0.0685 \text{ mol}$$
$$O_2 = \text{mol } O_2 \text{ entering} - \text{mol } O_2 \text{ reacted}$$
$$= (644 \text{ mol}_{\text{air}}) \left(0.21 \frac{\text{mol}_{O_2}}{\text{mol}_{\text{air}}} \right) - 104 \text{ mol}_{O_2}$$
$$= 31.2 \text{ mol}$$
$$N_2 = \text{mol } N_2 \text{ entering}$$
$$= (644 \text{ mol}_{\text{air}}) \left(0.79 \frac{\text{mol}_{N_2}}{\text{mol}_{\text{air}}} \right)$$
$$= 509 \text{ mol}$$
$$\text{total} = 72.7 \text{ mol} + 62.4 \text{ mol} + 0.0685 \text{ mol}$$
$$+ 31.2 \text{ mol} + 509 \text{ mol}$$
$$= 675 \text{ mol}$$
$$\frac{\text{mol}}{\text{fraction}} = \frac{\text{mol } SO_2}{\text{total mol}} = \frac{0.0685 \text{ mol}}{675 \text{ mol}}$$
$$= 1.01 \times 10^{-4}$$

The answer is B.

Problem 39

The proper placement of a pressure-relief valve and a rupture disc combination on a vessel is

(A) in series, with the rupture disc outside
(B) in series, with the pressure-relief valve outside
(C) in parallel
(D) either (A) or (B)

Solution

Opening the pressure-relief valve is the preferred mode of pressure release because it can be reset, as opposed to the rupture disc, which must be replaced. Therefore, the devices should be placed in parallel, sized so that the pressure-relief valve will open at a lower pressure and the rupture disc will burst at a higher pressure should the valve fail to vent quickly enough. Redundant systems improve the probability of avoiding catastrophic failure of the vessel.

The answer is C.

PROCESS CONTROL

Problems 40–42 are based on the following information and illustration.

A controlled process is shown in block diagram form in the following illustration. G_p, G_c, G_m, G_n, and G_d represent the Laplace domain transfer function of the controlled process, the controller, the measurement device, the measurement noise in the feedback signal, and the effect of the measured output disturbance d on the process output y. r is the setpoint signal. u is the output of the controller and the input to the process.

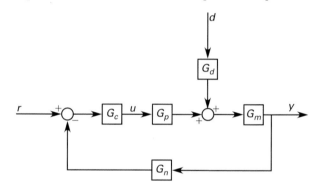

Problem 40

What is the closed-loop transfer function describing the effect of d and r on y?

(A) $y = \left(\dfrac{G_m G_d}{1 + G_m G_p G_c G_n} \right) d$
$\qquad + \left(\dfrac{G_m G_p G_c}{1 + G_m G_p G_c G_n} \right) r$

(B) $y = G_m G_d d + G_m G_p G_c (r - G_n)$

(C) $y = G_m G_d d + \left(\dfrac{G_m G_p G_c}{1 + G_m G_p G_c} \right) r$

(D) $y = \left(\dfrac{G_m G_d}{1 + G_m} \right) d + \left(\dfrac{G_m G_p G_c}{1 + G_n} \right) r$

Solution

The relations can be constructed in pieces.

$$y = G_m(G_d d + G_p u)$$
$$u = G_c(r - G_n y)$$

Combining the two equations gives

$$y = G_m(G_d d + G_p G_c(r - G_n y))$$

$$G_m G_p G_c G_n y + y = G_m G_d d + G_m G_p G_c r$$

$$y = \left(\frac{G_m G_d}{1 + G_m G_p G_c G_n}\right) d + \left(\frac{G_m G_p G_c}{1 + G_m G_p G_c G_n}\right) r$$

The answer is A.

Problem 41

If, in the process in the illustration, $G_n = 1$, $G_d = 0$, $G_m = s + 1$, $G_p = 1/(2s + 1)(s - 1)$, and $G_c = K$, what values of K will stabilize the controlled process?

(A) $K < 1$
(B) $K > 1$
(C) $K > 0.75$
(D) $K \geq 0$

Solution
The closed-loop transfer function is as expressed in Prob. 40.

$$y = \left(\frac{G_m G_d}{1 + G_m G_p G_c G_n}\right) d + \left(\frac{G_m G_p G_c}{1 + G_m G_p G_c G_n}\right) r$$

$$= \left(\frac{(1)(0)}{1 + (s+1)\left(\dfrac{1}{(2s+1)(s-1)}\right)(K)(1)}\right) d$$

$$+ \left(\frac{(s+1)\left(\dfrac{1}{(2s+1)(s-1)}\right)(K)}{1 + (s+1)\left(\dfrac{1}{(2s+1)(s-1)}\right)(K)(1)}\right) r$$

$$= \left(\frac{K(s+1)}{2s^2 + (K-1)s + (K-1)}\right) r$$

The controlled process is stable if the poles of the closed-loop transfer are in the left half of the Laplace plane; that is, if the real part of the roots of the denominator of the closed-loop transfer function are less than 0.

The roots of the denominator of the closed-loop transfer function are

$$s = \frac{-(K-1) \pm \sqrt{K^2 - 2K + 1 - (8)(K-1)}}{4}$$

$$= \frac{1 - K}{4} \pm \frac{\sqrt{K^2 - 10K + 9}}{4}$$

Borderline stability will occur when $s = 0$.

$$0 = \frac{1 - K}{4} \pm \frac{\sqrt{K^2 - 10K + 9}}{4}$$

$$(1 - K)^2 = K^2 - 10K + 9$$

$$1 - 2K + K^2 = K^2 - 10K + 9$$

$$K = 1$$

As K increases above 1, the roots of the polynomial remain positive, so the process remains stable for $K > 1$.

The answer is B.

Problem 42

Refer to the controlled process of the following illustration.

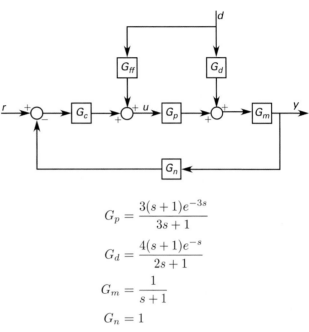

$$G_p = \frac{3(s+1)e^{-3s}}{3s+1}$$

$$G_d = \frac{4(s+1)e^{-s}}{2s+1}$$

$$G_m = \frac{1}{s+1}$$

$$G_n = 1$$

Design a realizable feedforward controller, G_{ff}, that transforms the measured disturbance, d, into a signal that, when added to the feedback controller output, u, counteracts the effect of the disturbance, d, on the output, y.

(A) $G_{ff} = \dfrac{(24s)(s+1)e^{-s}}{(2s+1)(3s+1)}$

(B) $G_{ff} = \left(-\dfrac{4}{3}\right)\left(\dfrac{3s+1}{2s+1}\right)$

(C) $G_{ff} = \left(-\dfrac{4}{3}\right)\left(\dfrac{(3s+1)e^{2s}}{2s+1}\right)$

(D) $G_{ff} = \left(-\dfrac{4}{3}\right)\left(\dfrac{(3s+1)e^{-2s}}{2s+1}\right)$

Solution
The effect of the disturbance, d, through the feedforward controller G_{ff} adds an additional term to the first half of the closed-loop transfer function of Prob. 40.

$$y = \left(\frac{G_m G_d}{1 + G_m G_p G_c G_n} + \frac{G_m G_p G_{ff}}{1 + G_m G_p G_c G_n}\right) d$$

$$+ \left(\frac{G_m G_p G_c}{1 + G_m G_p G_c G_n}\right) r$$

To eliminate the effect of d on y, the sum of the two bracketed terms must be 0.

$$G_m G_d = -G_m G_p G_{ff}$$

$$G_{ff} = -\frac{G_m G_d}{G_m G_p}$$

$$= -\frac{G_d}{G_p}$$

Substituting,

$$G_{ff} = -\frac{G_d}{G_p}$$

$$= \left(-\frac{4(s+1)e^{-s}}{2s+1}\right)\left(\frac{3s+1}{3(s+1)e^{-3s}}\right)$$

$$= \left(-\frac{4}{3}\right)\left(\frac{(3s+1)e^{2s}}{2s+1}\right)$$

This is the theoretically perfect feedforward controller. However, the factor e^{2s} implies that this controller must know the disturbance two units of time in the future. This is not physically realizable. The best physically realizable feedforward controller is, therefore,

$$G_{ff} = \left(-\frac{4}{3}\right)\left(\frac{3s+1}{2s+1}\right)$$

The answer is B.

PROCESS DESIGN AND ECONOMICS OPTIMIZATION

Problem 43

A tank is needed for the forseeable future. Four options meet the project requirements. The probability of failure in any year is equal to $1/n$ where n is the average lifetime of the vessel. The cost of such a failure would be $17,700 for any tank.

tank	average lifetime (yr)	cost ($)	annual maintenance ($)
1	10	20,300	1010
2	16	28,360	1060
3	25	36,400	1080
4	60	45,080	1130

Which tank should be purchased? Use a 10% interest rate.

(A) tank 1
(B) tank 2
(C) tank 3
(D) tank 4

Solution

The four options have different lifetimes, so it is necessary to compare them on an equivalent uniform annual cost (EUAC) basis. The annual cost sums the annualized initial investment, the annual maintenance cost, and the cost of failure multiplied by the probability of failure.

$$\text{EUAC} = P(A/P, 10\%, n) + M + \frac{1}{n}F$$

$$\text{EUAC}_1 = (\$20{,}300)(0.1627) + \$1010$$
$$+ \left(\frac{1}{10\text{ yr}}\right)(\$17{,}700)$$
$$= \$6083/\text{yr}$$

$$\text{EUAC}_2 = (\$28{,}360)(0.1278) + \$1060$$
$$+ \left(\frac{1}{16\text{ yr}}\right)(\$17{,}700)$$
$$= \$5791/\text{yr}$$

$$\text{EUAC}_3 = (\$36{,}400)(0.1102) + \$1080$$
$$+ \left(\frac{1}{25\text{ yr}}\right)(\$17{,}700)$$
$$= \$5799/\text{yr}$$

$$\text{EUAC}_4 = (\$45{,}080)(0.1003) + \$1130$$
$$+ \left(\frac{1}{60\text{ yr}}\right)(\$17{,}700)$$
$$= \$5947/\text{yr}$$

Tank 2 has the lowest annualized cost.

The answer is B.

Problems 44 and 45 are based on the following information.

A new piece of equipment costs $180,000 and will be depreciated with straight-line depreciation. The depreciation cost may not exceed $10,000 in any year. The scrap value of the equipment is $20,000.

Problem 44

What is the minimum service life that will justify the expense?

(A) 14 yr
(B) 16 yr
(C) 18 yr
(D) 20 yr

Solution

Straight-line depreciation divides the net cost (purchase price − salvage value) by the lifetime of the equipment.

$$D_j = \frac{C - S_n}{n}$$

Therefore,

$$n = \frac{C - S_n}{D_j}$$

$$= \frac{\$180{,}000 - \$20{,}000}{\dfrac{\$10{,}000}{1 \text{ yr}}}$$

$$= 16 \text{ yr}$$

The answer is B.

Problem 45

The book value of the equipment after 10 yr is

(A) $20,000
(B) $60,000
(C) $80,000
(D) $100,000

Solution

The book value is the worth of an item after subtracting cumulative depreciation.

$$\text{BV} = \text{initial cost} - \sum D_j$$

$$= \$180{,}000 - (10 \text{ yr})\left(\frac{\$10{,}000}{1 \text{ yr}}\right)$$

$$= \$80{,}000$$

The answer is C.

Problems 46–48 are based on the following information.

A chemical is produced in a batch operation in which the cycle time is related to the batch size (P_b in kg/batch). The charge and discharge time is $0.25 + 0.06P_b^{0.3}$ in h/batch. The batch running time is $2.2P_b^{0.3}$ in h/batch.

A batch size of 1000 kg–10 000 kg is dictated by equipment constraints. Shutdown time between batches for maintenance and equipment repair is 5.0% of the operating cycle time (including discharge and charge time). The operating costs are $20/h during discharge and charge and $50/h during the run. The annual equipment costs vary with batch size as

$$\text{equipment cost} = \$420P_b^{0.75}/\text{yr}$$

The annual production of 1.50 million kilograms incurs fixed overhead and material costs of $752,000/yr. The product sells for $1.03/kg.

Problem 46

The batch size that minimizes annual costs is most nearly

(A) 6000 kg
(B) 7000 kg
(C) 10 000 kg
(D) 12 000 kg

Solution

The annual costs are summed as follows.

$$C_y = ((\text{batch operation time})(\text{hourly cost}))\left(\frac{\text{batches}}{\text{yr}}\right)$$

$$+ \text{annual equipment costs} + \text{annual fixed costs}$$

$$= \begin{pmatrix} (0.25 + 0.06P_b^{0.3})\left(\dfrac{\$20}{1 \text{ h}}\right) \\[2mm] + (2.2P_b^{0.3})\left(\dfrac{\$50}{1 \text{ h}}\right) \end{pmatrix} \begin{pmatrix} 1.50 \times 10^6 \, \dfrac{\text{kg}}{\text{yr}} \\[2mm] \overline{P_b} \end{pmatrix}$$

$$+ 420P_b^{0.75} + \frac{\$752{,}000}{1 \text{ yr}}$$

$$= (7.50 \times 10^6)P_b^{-1} + (1.67 \times 10^8)P_b^{-0.7}$$

$$+ 420P_b^{0.75} + \$752{,}000$$

The annual cost is at a minimum where the derivative of the annual cost with respect to batch size is zero.

$$\frac{dC_y}{dP_b} = (-7.50 \times 10^6)P_b^{-2} - (1.17 \times 10^8)P_b^{-1.7}$$

$$+ 315P_b^{-0.25}$$

$$= 0$$

The equation can be solved iteratively using Newton's method of root extraction. First, an estimate is made for P_b. A new estimate is calculated as follows.

$$P_b^{j+1} = P_b^j - \frac{\dfrac{dC_y}{dP_b}}{\dfrac{d^2 C_y}{dP_b^2}}\bigg|_{P_b = P_b^j}$$

The second derivative of the cost equation is

$$\frac{d^2 C_y}{dP_b^2} = (1.50 \times 10^7)P_b^{-3} + (1.99 \times 10^8)P_b^{-2.7}$$

$$- 78.8P_b^{-1.25}$$

Iterations are shown as follows.

batch size, P_b (kg)	$\dfrac{dC_y}{dP_b}$	$\dfrac{d^2 C_y}{dP_b^2}$	$\dfrac{\frac{dC_y}{dP_b}}{\frac{d^2 C_y}{dP_b^2}}$
5000	−23.1	0.0187	−1235
6235	−6.14	0.009 93	−618
6853	−0.791	0.007 53	−105
6958	−0.0180	0.007 20	−2.5

A batch size of 6960 kg (7000 kg) minimizes the annual cost.

The answer is B.

Problem 47

The number of minimal-cost batches produced each week (168 h) is most nearly

    (A) 4.1 batches/wk
    (B) 4.4 batches/wk
    (C) 4.9 batches/wk
    (D) 5.2 batches/wk

Solution

The cycle time, including the shutdown between batches, is

cycle time = charge and discharge time
              + running time + shutdown time

$$= \left(0.25 + 0.06 P_b^{0.3} \; \frac{h}{batch} + 2.2 P_b^{0.3} \; \frac{h}{batch}\right)$$
$$\times (1.05)$$
$$= \left(0.25 + 2.26 P_b^{0.3} \; \frac{h}{batch}\right)(1.05)$$

The number of batches per week is then

$$\frac{168 \; \frac{h}{wk}}{\left(0.25 + 2.26 P_b^{0.3} \; \frac{h}{batch}\right)(1.05)}$$

$$= \frac{168 \; \frac{h}{wk}}{\left(0.25 + \left(2.26 \; \frac{h}{batch}\right)(6960)^{0.3} \; \frac{h}{batch}\right)(1.05)}$$

$$= 4.94 \; batches/wk \quad (4.9 \; batches/wk)$$

The answer is C.

Problem 48

The batch size at which the production breaks even financially is most nearly

    (A) 1000 kg
    (B) 2900 kg
    (C) 4500 kg
    (D) 10 000 kg

Solution

The production breaks even when production costs equal the selling price.

$$\frac{price}{kg} = \frac{\frac{operation\;cost}{batch}}{\frac{kg}{batch}} + \frac{\frac{equipment\;and\;material\;cost}{yr}}{\frac{kg}{yr}}$$

$$1.03 \; \frac{\$}{kg} = \frac{(0.25 + 0.06 P_b^{0.3})\left(20 \; \frac{\$}{h}\right)}{P_b}$$
$$+ \frac{(2.2 P_b^{0.3})\left(50 \; \frac{\$}{h}\right)}{P_b}$$
$$+ \frac{(420 P_b^{0.75}) + 7.52 \times 10^5 \; \frac{\$}{yr}}{1.50 \times 10^6 \; \frac{kg}{yr}}$$

$$= 5.00 P_b^{-1} + 111 P_b^{-0.7} + (2.80 \times 10^{-4}) P_b^{0.75}$$
$$+ 0.501$$
$$0 = 5.00 P_b^{-1} + 111 P_b^{-0.7} + (2.80 \times 10^{-4}) P_b^{0.75}$$
$$- 0.529$$
$$= f(P_b)$$

The equation is solved for P_b using Newton's method as before.

$$P_b^{j+1} = P_b^j - \frac{f(P_b)}{f'(P_b)}\bigg|_j$$

$$f'(P_b) = -5.00 P_b^{-2} - 77.7 P_b^{-1.7}$$
$$+ (2.10 \times 10^{-4}) P_b^{-0.25}$$

Iterations are shown as follows.

batch size, P_b (kg)	$f(P_b)$	$f'(P_b)$	$\frac{f(P_b)}{f'(P_b)}$
2000	0.100	-1.60×10^{-4}	-625
2625	0.0243	-9.10×10^{-5}	-267
2892	0.00241	-7.34×10^{-5}	-32.8
2925	1.94×10^{-5}	-7.16×10^{-5}	-0.271

The batch size at the break-even point is 2930 kg (2900 kg).

The answer is B.

CHEMISTRY

Problem 49

In what type of reaction would a single compound produce two or more substances?

   (A) displacement
   (B) decomposition
   (C) polymerization
   (D) reduction

Solution

Decomposition occurs when a single compound produces two or more substances. Displacement occurs when an element reacts with a compound to displace an element from that compound. Polymerization is a joining together of compounds. Reduction occurs when a substance gains an electron.

The answer is B.

Problem 50

The reaction of halides with imide salts is important in the production of pure amines. This process is called the

   (A) Williamson synthesis
   (B) Strecker synthesis
   (C) Beckmann rearrangement
   (D) Gabriel synthesis

Solution

The Gabriel synthesis refers to the reaction of halides with imide salts. The Williamson synthesis is used in the preparation of ethers. The Strecker synthesis is a route to the formation of amino acids. The Beckmann rearrangement is a rearrangement of the oximes of ketones.

The answer is D.

Problem 51

What is most nearly the theoretical yield of bromobenzene when 40 g of benzene and 60 g of bromine are used in the following reaction?

$$C_6H_6 + Br_2 \rightarrow C_6H_5Br + HBr$$

   (A) 30 g
   (B) 41 g
   (C) 59 g
   (D) 74 g

Solution

1 mol of C_6H_6 reacts to form 1 mol of C_6H_5Br. 1 mol of Br_2 reacts to form 1 mol of C_6H_5Br. The molecular weights of these compounds are as follows.

$$MW_{C_6H_6} = 78 \text{ g/mol}$$
$$MW_{Br_2} = 159.8 \text{ g/mol}$$
$$MW_{C_6H_5Br} = 156.9 \text{ g/mol}$$

The number of moles of C_6H_6 and Br_2 are determined as follows.

$$N_{C_6H_6} = \frac{m_{benzene}}{MW_{C_6H_6}} = \frac{40 \text{ g}}{78 \frac{g}{mol}}$$
$$= 0.513 \text{ mol}$$
$$N_{Br_2} = \frac{m_{bromine}}{MW_{Br_2}} = \frac{60 \text{ g}}{159.8 \frac{g}{mol}}$$
$$= 0.375 \text{ mol}$$

Bromine is the limiting reactant, since there are less moles of bromine to react with benzene (0.375 mol bromine versus 0.513 mol of benzene). Therefore the theoretical yield of bromobenzene is

$$m_{theoretical\ yield} = N_{Br_2}MW_{C_6H_5Br}$$
$$= (0.375 \text{ mol})\left(156.9 \frac{g}{mol}\right)$$
$$= 58.8 \text{ g} \quad (59 \text{ g})$$

The answer is C.

Problem 52

What is the correct expression of K_{eq} for the complete combustion of ethane?

  (A) $K_{eq} = \dfrac{(CO_2)(H_2O)}{(C_2H_6)O_2}$

  (B) $K_{eq} = \dfrac{(CO_2)^2(H_2O)^3}{(C_2H_6)(O_2)^3}$

  (C) $K_{eq} = \dfrac{(CO_2)^2(H_2O)^3}{C_2H_6}$

  (D) $K_{eq} = \dfrac{(CO_2)^4(H_2O)^6}{(C_2H_6)^2(O_2)^7}$

Solution

The correct answer is derived when products and reactants are raised to their molecular prefixes. The molecular prefixes are

$$2\,C_2H_6 + 7\,O_2 \rightarrow 4\,CO_2 + 6\,H_2O$$

The answer is D.

Problem 53

A 5 m ethanol solution of a nonvolatile solute has a 78.4°C normal boiling point and a 1.22°C/m boiling-point constant. What is the boiling point of the solution?

(A) 83.4°C
(B) 84.5°C
(C) 85.6°C
(D) 86.7°C

Solution

The boiling point is determined from the sum of the boiling-point elevation and normal boiling point.

$$BP = \Delta T_b + NBP$$

To determine the boiling-point elevation, use the following formula.

$$\Delta T_b = K_b m = \left(1.22 \ \frac{°C}{m}\right)(5 \ m)$$

$$= 6.1°C$$

Substituting,

$$BP = 6.1°C + 78.4°C$$
$$= 84.5°C$$

The answer is B.

Problem 54

It is desired to produce a salt solution with a molality of 0.30 mol/kg. Approximately how much water is required if the amount of NaCl is 155 g?

(A) 8.8 kg
(B) 11 kg
(C) 430 kg
(D) 520 kg

Solution

Molality, m, is determined from the number of moles of solute, N, per mass of solvent, also m, in kilograms. So in this case,

$$m = \frac{N_{NaCl}}{m_{H_2O}}$$

Substituting, and converting 155 g to moles,

$$0.30 \ \frac{mol}{kg} = \frac{(155 \ g)\left(\frac{1 \ mol}{58.5 \ g}\right)}{m_{H_2O}}$$

$$= \frac{265 \ mol}{m_{H_2O}}$$

Rearranging to solve for the mass of water needed,

$$m_{H_2O} = \frac{2.65 \ mol}{0.30 \ \frac{mol}{kg}}$$

$$= 8.83 \ kg \quad (8.8 \ kg)$$

The answer is A.

FLUID DYNAMICS

Problem 55

The flow rate of benzene is being measured by a venturi meter in a pipe with a 75 mm internal diameter. The venturi meter has a throat diameter of 50 mm. A mercury manometer shows a differential of 15 mm. The venturi meter has a flow coefficient of 0.98. Under flowing conditions, benzene has a density of 820 kg/m³ and a viscosity of 1 cP. What is most nearly the flow rate of benzene in the pipe?

(A) 0.0050 m³/h
(B) 0.28 m³/h
(C) 17 m³/h
(D) 74 m³/h

Solution

Flow rate is determined from

$$Q = v_t A_t$$

The velocity of benzene at the throat, v_t, is

$$v_t = C \sqrt{\frac{2g\Delta H}{1 - \left(\frac{D_t}{D_p}\right)^4}}$$

The Newton's law proportionality factor, g_c, is 9.81 m/sec². The differential head, ΔH, is determined from

$$\Delta H = h_m \left(\frac{S_m}{S_{benzene}} - 1\right)$$

The specific gravity of mercury is 13.6, and the specific gravity of benzene is 0.820. Substituting,

$$\Delta H = (15 \ mm)\left(\frac{13.6}{0.82} - 1\right)$$

$$= 233.8 \ mm \quad (0.234 \ m)$$

Now, solving for the velocity of benzene at the throat,

$$v_T = 0.98 \sqrt{\frac{(2)\left(9.81 \ \frac{m}{s^2}\right)(0.234 \ m)}{1 - \left(\frac{50 \ mm}{75 \ mm}\right)^4}}$$

$$= 2.34 \ m/s$$

The throat area, A_T, is determined from

$$A_T = \frac{\pi D_t^2}{4}$$

Convert the 50 mm throat diameter to meters, and solve.

$$A_T = \frac{\pi (0.050 \text{ m})^2}{4}$$
$$= 0.001\,96 \text{ m}^2$$

Substituting to solve for the flow rate,

$$Q = \left(2.34 \frac{\text{m}}{\text{s}}\right) (0.001\,96 \text{ m}^2)$$
$$= \left(0.004\,59 \frac{\text{m}^3}{\text{s}}\right) \left(3600 \frac{\text{s}}{\text{h}}\right)$$
$$= 16.5 \text{ m}^3/\text{h} \quad (17 \text{ m}^3/\text{h})$$

The answer is C.

Problems 56–60 are based on the following information.

Gasoline is being pumped through a pipe with a 150 mm internal diameter at a rate of 150 m^3/h over a distance of 525 m. The pipe has five long-radius (LR) elbows, two 45° elbows, and two gate valves. The gasoline has a specific gravity of 0.67 and a viscosity of 0.80 cP. The friction factor can be estimated from the following equation.

$$f = \left(1.8 \log \left(\frac{\text{Re}}{7}\right)\right)^{-2}$$

Problem 56
What is most nearly the velocity in the pipe?

(A) 0.60 m/s
(B) 2.4 m/s
(C) 5.8 m/s
(D) 10 m/s

Solution
The velocity in the pipe is determined from

$$v = \frac{Q}{A}$$

Convert the flow rate to cubic meters per second.

$$Q = \left(150 \frac{\text{m}^3}{\text{h}}\right) \left(\frac{1 \text{ h}}{3600 \text{ s}}\right)$$
$$= 0.042 \text{ m}^3/\text{s}$$

The pipe area is determined from

$$A = \frac{\pi D^2}{4}$$

Converting the 150 mm pipe internal diameter to meters and solving,

$$A = \frac{\pi (0.150 \text{ m})^2}{4}$$
$$= 0.0177 \text{ m}^2$$

Substituting to solve for the velocity in the pipe,

$$v = \frac{0.042 \frac{\text{m}^3}{\text{s}}}{0.0177 \text{ m}^2}$$
$$= 2.37 \text{ m/s} \quad (2.4 \text{ m/s})$$

The answer is B.

Problem 57
What is the Reynolds number for this flow scheme, if the velocity is 4.1 m/s?

(A) 1.5×10^3
(B) 1.5×10^4
(C) 2.5×10^5
(D) 5.1×10^5

Solution
The Reynolds number is determined from

$$\text{Re} = \frac{Dv\rho}{\mu}$$

The diameter is given as 150 mm, or 0.150 m.

The specific gravity, which is given as 0.67, is converted to specific density as follows.

$$\rho = (0.67) \left(1000 \frac{\text{kg}}{\text{m}^3}\right)$$
$$= 670 \text{ kg/m}^3$$

Converting the given viscosity to kg/m·s,

$$\mu = (0.8 \text{ cP}) \left(0.001 \frac{\frac{\text{kg}}{\text{m·s}}}{\text{cP}}\right)$$
$$= 0.0008 \text{ kg/m·s}$$

The Reynolds number can now be determined.

$$\text{Re} = \frac{(0.150 \text{ m}) \left(4.1 \frac{\text{m}}{\text{s}}\right) \left(670 \frac{\text{kg}}{\text{m}^3}\right)}{0.0008 \frac{\text{kg}}{\text{m·s}}}$$
$$= 515\,062 \quad (5.1 \times 10^5)$$

The answer is D.

Problem 58

What is the friction factor for this flow scheme if the Reynolds number can be taken as 2×10^4?

(A) 0.0030
(B) 0.010
(C) 0.025
(D) 0.032

Solution

Use the Reynolds number to find the friction factor.

$$f = \left(1.8 \log \frac{\text{Re}}{7}\right)^{-2}$$
$$= \left(1.8 \log \left(\frac{2.0 \times 10^4}{7}\right)\right)^{-2}$$
$$= 0.025$$

The answer is C.

Problem 59

The length-to-diameter ratio, L/D, values for the pipe fittings are as follows.

$$(L/D)_{\text{LR elbows}} = 20$$
$$(L/D)_{45^\circ \text{ elbows}} = 16$$
$$(L/D)_{\text{gate valves}} = 13$$

What is the equivalent length of the pipe fittings?

(A) 8.8 m
(B) 24 m
(C) 31 m
(D) 35 m

Solution

The length of each fitting is determined from

$$L = D \frac{L}{D}$$

For each fitting, find this value, and multiply it by the number of fittings. Convert the pipe internal diameter, 150 mm, to meters such that D is 0.150 m.

For the LR elbows,

$$L_{\text{LR elbows}} = (0.150 \text{ m})(20)(5)$$
$$= 15 \text{ m}$$

For the 45° elbows,

$$L_{45^\circ \text{ elbows}} = (0.150 \text{ m})(16)(2)$$
$$= 4.8 \text{ m}$$

For the gate valves,

$$L_{\text{gate valves}} = (0.150 \text{ m})(13)(2)$$
$$= 3.9 \text{ m}$$

The equivalent length of the pipe fittings is the sum of the length values.

$$L_{\text{eq}} = L_{\text{LR elbows}} + L_{45^\circ \text{ elbows}} + L_{\text{gate valves}}$$
$$= 15 \text{ m} + 4.8 \text{ m} + 3.9 \text{ m}$$
$$= 23.7 \text{ m} \quad (24 \text{ m})$$

The answer is B.

Problem 60

What is the pressure drop through this section of pipe, if the friction factor is 0.018, the velocity is 3.0 m/s, and the equivalent length of pipe and fittings is 650 m?

(A) 15 m
(B) 20 m
(C) 25 m
(D) 36 m

Solution

The pressure drop, or head loss due to pipe flow, through this section of pipe is determined from

$$h_f = \frac{fLv^2}{2Dg}$$

The friction factor is 0.018, the velocity in the pipe is 3.0 m/s, and the equivalent length of pipe is 650 m.

Convert the 150 mm pipe internal diameter to meters, such that D is 0.150 m.

The acceleration of gravity is 9.81 m/s².

Substitute these values in the following equation to determine the pressure drop.

$$h_f = \frac{(0.018)(650 \text{ m})\left(3.0 \frac{\text{m}}{\text{s}}\right)^2}{(2)(0.150 \text{ m})\left(9.81 \frac{\text{m}}{\text{s}^2}\right)}$$
$$= 35.8 \text{ m} \quad (36 \text{ m})$$

The answer is D.

Practice Exam 1

1. The half-life of cyclobutane at 750K is 194 s. How long will it take for 25% of an original charge to decompose in this first-order reaction?

- (A) 81 s
- (B) 97 s
- (C) 190 s
- (D) 390 s

Problems 2–3 are based on the following information.

Ethyl acetate is produced from ethanol and acetic acid in a reversible second-order condensation reaction.

$$C_2H_5OH + CH_3COOH \underset{k_2}{\overset{k_1}{\rightleftharpoons}} CH_3COOC_2H_5 + H_2O$$

An 80 wt% ethanol stream and a 75 wt% acetic acid stream, both at 25°C, flow into a 10.0 m³ CSTR, where the reaction proceeds isothermally at 70°C to 35% conversion of the limiting reagent. The two streams enter at the same volumetric flow rate. Assume constant density for all components and mixtures of 1 g/cm³. Neglect any heats of mixing.

At 70°C,
$$k_1 = 7.93 \times 10^{-6} \text{ L/mol·s}$$
$$k_2 = 2.71 \times 10^{-6} \text{ L/mol·s}$$

At 25°C,
$$\text{heat of reaction} = 11.3 \text{ kJ/mol}$$

In the temperature range of 0–100°C,
$$\text{specific heat of ethanol} = 142 \text{ J/mol·°C}$$
$$\text{specific heat of acetic acid} = 183 \text{ J/mol·°C}$$
$$\text{specific heat of ethyl acetate} = 242 \text{ J/mol·°C}$$
$$\text{specific heat of water} = 75.4 \text{ J/mol·°C}$$

2. The concentrations of species in the aqueous effluent stream are most nearly

- (A) 4.64 mol/L ethanol, 2.19 mol/L acetic acid, and 4.06 mol/L ethyl acetate
- (B) 6.51 mol/L ethanol, 4.06 mol/L acetic acid, and 2.19 mol/L ethyl acetate
- (C) 9.27 mol/L ethanol, 4.38 mol/L acetic acid, and 8.13 mol/L ethyl acetate
- (D) 13.0 mol/L ethanol, 8.13 mol/L acetic acid, and 4.38 mol/L ethyl acetate

3. The volumetric flow rate of each feed stream is

- (A) 0.28 L/s
- (B) 0.47 L/s
- (C) 0.56 L/s
- (D) 1.9 L/s

Problems 4–6 are based on the following information.

Nitrogen dioxide and carbon monoxide react to form less hazardous gases in a second-order reaction.

$$NO_2 + CO \longrightarrow NO + CO_2$$

A mixture containing 78.2 vol% nitrogen, 13.4 vol% carbon monoxide, and 8.4 vol% nitrogen dioxide is charged into an isothermal batch reactor held at 400°C. The reaction proceeds to 99% conversion of NO_2 to NO. The initial pressure is 101 kPa. Assume the ideal gas law applies. The data for reaction rates measured over a range of temperatures are shown in the table.

temperature (°C)	rate constant $\left(\dfrac{\text{L}}{\text{mol·s}}\right)$
350	0.0765
450	2.74
550	41.1

4. The reaction activation energy is most nearly

- (A) 0.262 kJ/mol
- (B) 16.1 kJ/mol
- (C) 49.9 kJ/mol
- (D) 134 kJ/mol

5. Determine the approximate gas mixture composition at 99% conversion.

- (A) 77.24% N_2, 1.32% CO, 0.08% NO_2, 13.14% CO_2, 8.22% NO
- (B) 78.20% N_2, 1.27% CO, 0.08% NO_2, 12.57% CO_2, 7.88% NO
- (C) 78.20% N_2, 5.08% CO, 0.08% NO_2, 8.32% CO_2, 8.32% NO
- (D) 78.20% N_2, 8.32% CO, 8.32% NO_2, 5.08% CO_2, 0.08% NO

6. How much time is required to reach 99% conversion of NO_2?

(A) 139 s
(B) 47.7 min
(C) 128 min
(D) 34.7 h

7. The diffusivity of ammonia through air at 273K and 101 kPa is 1.98×10^{-5} m^2/s. What is the diffusivity at 300K and 150 kPa?

(A) 1.47×10^{-5} m^2/s
(B) 1.54×10^{-5} m^2/s
(C) 2.55×10^{-5} m^2/s
(D) 2.67×10^{-5} m^2/s

Problems 8–9 are based on the following information.

Acetic acid is to be separated from water in a continuous distillation column with a total condenser and partial reboiler to a final concentration of 95 mol% acetic acid. The other, water-rich product stream may contain up to 10 mol% acetic acid. The column will process a 60 wt% acetic acid-saturated liquid feedstream at 35 380 kg/h. The column will be run with a reflux ratio that is 80% above the minimum. Equilibrium data for acetic acid and water at 101.3 kPa total pressure are given as follows.

mole fraction of water	
liquid	vapor
0.0000	0.0000
0.1881	0.3063
0.3084	0.4467
0.4498	0.5973
0.5195	0.6580
0.5824	0.7112
0.6750	0.7797
0.7261	0.8239
0.7951	0.8671
0.8556	0.9042
0.8787	0.9186
0.9134	0.9409
0.9578	0.9708
1.0000	1.0000

8. What is the concentrated acetic acid production rate?

(A) 282 kg/h
(B) 11 900 kg/h
(C) 15 200 kg/h
(D) 16 300 kg/h

9. Determine the minimum reflux ratio for the column.

(A) 1.10
(B) 1.70
(C) 2.57
(D) 3.78

Problems 10–12 are based on the following information.

A recycled furfural stream is used to extract benzene from a cyclohexane product stream. The product stream flows at 100 kg/h and contains 10 wt% benzene. The solvent stream flows at 150 kg/h and contains 0.01 wt% benzene. The distribution coefficient for benzene in the two phases is 0.680, based on the weight fraction of solute. It is desired to reduce the benzene content in the product stream to less than 1 wt% benzene. Assume cyclohexane and furfural are immiscible.

10. The raffinate composition in a single-stage extraction column is most nearly

(A) 5.00 wt% benzene and 95.00 wt% cyclohexane
(B) 6.80 wt% benzene and 93.20 wt% cyclohexane
(C) 6.80 wt% benzene and 93.20 wt% furfural
(D) 7.35 wt% benzene and 92.65 wt% furfural

11. The raffinate flow rate in a single-stage extraction is most nearly

(A) 94.7 kg/h
(B) 96.6 kg/h
(C) 155 kg/h
(D) 157 kg/h

12. The extract composition from the last stage of the countercurrent extractor is most nearly

(A) 3.4 wt% benzene and 96.6 wt% furfural
(B) 4.0 wt% benzene and 96.0 wt% cyclohexane
(C) 5.7 wt% benzene and 94.3 wt% furfural
(D) 6.8 wt% benzene and 93.2 wt% cyclohexane

Problems 13–15 are based on the following information.

An Orsat analysis determined a stack gas was composed of 3.0 mol% carbon monoxide, 10.6 mol% carbon dioxide, 6.0 mol% oxygen, and 80.4 mol% nitrogen. Dry air enters the combustion chamber.

13. The amount of carbon burned is most nearly

(A) 9.90 mol/100 mol stack gas
(B) 10.6 mol/100 mol stack gas
(C) 12.7 mol/100 mol stack gas
(D) 13.6 mol/100 mol stack gas

14. The amount of air used is most nearly

(A) 75.4 mol/100 mol stack gas
(B) 80.4 mol/100 mol stack gas
(C) 95.4 mol/100 mol stack gas
(D) 102 mol/100 mol stack gas

15. The excess air is most nearly

(A) 4.5%
(B) 6.0%
(C) 21%
(D) 27%

Problems 16–19 are based on the following information.

Natural gas is burned in a furnace at a rate of $45.0 \text{ m}^3/$ day. Fuel and air enter the furnace at 101.3 kPa and 25.0°C. Natural gas is composed mainly of methane, but may include up to 10.0 mol% ethane, 3.0 mol% propane, 1.0 mol% butane, and 0.5 mol% pentane. The typical natural gas makeup is tabulated with the following combustion data. Heat capacity data are also provided.

component	typical mol%	heat of combustion at 25°C (kJ/mol)
methane	94.40	−890.4
ethane	3.40	−1560
propane	0.60	−2220
butane	0.50	−2874
pentane	–	−3533
carbon dioxide	0.60	–
nitrogen	0.50	–

The heat capacity (in J/mol·°C) in terms of temperature T (in °C) is

$$(c_p)_{CO_2} = 36.80 + 1.406 \times 10^{-2}T$$
$$(c_p)_{H_2O} = 33.43 + 8.392 \times 10^{-3}T$$
$$(c_p)_{O_2} = 29.27 + 4.683 \times 10^{-3}T$$
$$(c_p)_{N_2} = 28.99 + 2.729 \times 10^{-3}T$$

16. The natural gas is to be burned with 100% excess air. Determine the air-flow rate based on the typical fuel composition.

(A) 90.0 m³/d
(B) 186 m³/d
(C) 388 m³/d
(D) 887 m³/d

17. What is the typical flue gas molar composition assuming complete combustion?

(A) 5.09% CO₂, 9.83% H₂O, 9.98% O₂, 75.1% N₂
(B) 7.37% CO₂, 14.3% H₂O, 4.93% O₂, 73.4% N₂
(C) 9.65% CO₂, 18.7% H₂O, 0.05% O₂, 71.6% N₂
(D) 20.3% CO₂, 39.5% H₂O, 40.1% O₂, 0.09% N₂

18. The walls of the furnace must be constructed of materials capable of withstanding the highest temperatures produced in the furnace. Assuming a constant fuel-flow rate, calculate the highest theoretical temperature produced in the furnace.

(A) 2130°C
(B) 2170°C
(C) 2320°C
(D) 2360°C

19. The flue gas-flow rate that occurs under the conditions posed in Prob. 18 is most nearly

(A) 3970 m³/d
(B) 4470 m³/d
(C) 6670 m³/d
(D) 7640 m³/d

20. A piece of equipment is purchased for $27,500. After the anticipated useful life of 10 yr, the scrap value will be $1200. The equipment is depreciated using the accelerated cost recovery system. Changes to the process make the equipment obsolete after only 3 yr. What is the book value 3 yr after purchase?

(A) $15,800
(B) $16,400
(C) $19,300
(D) $19,700

Problems 21 and 22 are based on the following information.

For an initial investment of $189,000, a new product line will bring in an income of $38,000/yr. After 15 yr, the equipment will have a salvage value of $63,000. Annual maintenance and raw material costs will total $3520. Re-tooling and upgrade every 5 yr will cost $7500. The effective annual interest rate is 12%.

21. What is the present value of the investment?

(A) $49,300
(B) $50,700
(C) $50,900
(D) $189,000

22. What is the present value taking into account a 3% inflation rate?

(A) $11,100
(B) $14,800
(C) $27,600
(D) $251,000

23. An old reactor requires $2000 annually in maintenance and repairs to keep up with production. The product line will be continued for 10 yr. Three alternatives to enable continued operation are proposed. An investment of $14,500 now will upgrade the equipment so it will last for 10 yr without continuing maintenance costs. Alternatively, $7000 now will correct critical problems, allowing operation to continue for 5 yr, at which time an additional $10,000 will be needed for the remaining 5 yr. Equivalent operation is also obtained with an initial investment of $10,000 and a $7000 upgrade in 5 yr. Company policy mandates at least an 8% annual rate of return on all investments. Which alternative should be selected?

(A) Make no investment.
(B) Invest $14,500 now.
(C) Invest $7000 now and $10,000 in 5 yr.
(D) Invest $10,000 now and $7000 in 5 yr.

24. An engineer invests $2000 in an Individual Retirement Account (IRA) each year during her 40 yr career. Assume the investment earns an 8% rate of return. What annual income can be drawn from the investment for the 30 yr following retirement?

(A) $2670
(B) $4560
(C) $17,300
(D) $46,000

25. A company wishes to purchase manufacturing equipment to produce a new product for the forseeable future. Four alternative investments are being considered. All four investments are for the same type of unit and yield the same quantity and quality product. Given a discount rate (interest rate) of 12%, which is the most economical option?

unit	A	B	C	D
life	10 yr	10 yr	15 yr	20 yr
cost	$20,000	$25,000	$30,000	$45,000
maintenance	$1800/yr	$1200/yr	$1200/yr	$500/yr

(A) Alternative A is most economical.
(B) Alternative B is most economical.
(C) Alternative C is most economical.
(D) Alternative D is most economical.

26. Consider the process in the following illustration. If $G_p = 4$ s, and $G_c = 1 + 1/s^n$, what must the value of n be to guarantee steady-state tracking of the setpoint r by the output y free of offset?

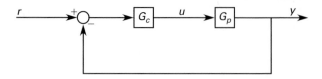

(A) 0
(B) 1
(C) 2
(D) 3

27. In the process of the illustration in Prob. 26, $G_p = 1/(2s+1)$ and $G_c = K/(s+2)$. Determine the value of K to achieve a critically damped response of the output y to a change in the setpoint r.

(A) −0.875
(B) 1.13
(C) 2.13
(D) 4.25

28. The response of the output of a process P to a step change in its input is depicted in the following illustration. If the process P is to be modeled using a first-order plus dead-time transfer function, that transfer function is most nearly

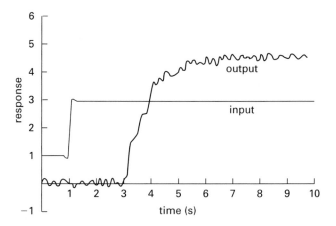

(A) $G_p = \dfrac{2.5e^{-2s}}{s+1}$

(B) $G_p = \dfrac{0.4e^{-3s}}{3s+1}$

(C) $G_p = \dfrac{0.4e^{-2s}}{2s+1}$

(D) $G_p = \dfrac{2.5e^{-3s}}{3s+1}$

Problems 29–31 are based on the following information.

Lubricating oil with a density of 860 kg/m^3 is pumped through 300 m of horizontal pipe at 4.44 m^3/h. The pipe inside diameter is 5.25 cm, and the pressure drop over the length of the pipe is 208 kPa.

29. The shear stress at the pipe wall is most nearly

(A) 0.46 Pa
(B) 0.91 Pa
(C) 4.55 Pa
(D) 9.10 Pa

30. The oil viscosity is most nearly

(A) 0.105 Pa·s
(B) 0.210 Pa·s
(C) 0.420 Pa·s
(D) 0.840 Pa·s

31. The fluid velocity 1 cm from the pipe wall is most nearly

(A) 0.392 m/s
(B) 0.702 m/s
(C) 1.57 m/s
(D) 2.81 m/s

32. A pump delivers water from a tank at ambient conditions to a wash tower downstream. It is found that the pump develops 25 m of total head and has a pumping power requirement of 13 kW. The pump efficiency is 65%. What is most nearly the volumetric flow rate of water delivered to the wash tower?

(A) 110 m^3/h
(B) 120 m^3/h
(C) 190 m^3/h
(D) 230 m^3/h

33. It is recommended that the velocity of a semi-abrasive fluid be limited to a maximum of 0.6 m/s. 250 m^3/h of this fluid must be pumped to a filter press 100 m away. The fluid's specific gravity is 1.03, and its viscosity is 1.5 cP. What commercially available pipe should be installed based on the internal diameter required for this service?

(A) 361 mm
(B) 387 mm
(C) 412 mm
(D) 438 mm

34. Ethylene gas is flowing through a nozzle at a mach number of 0.1. The gas temperature is 200°C. The specific heat ratio of ethylene is 1.28. What is the most nearly the velocity in the nozzle?

(A) 13 m/s
(B) 28 m/s
(C) 33 m/s
(D) 42 m/s

Problems 35 and 36 are based on the following information.

The bulk temperature of fluid in a stainless steel reactor (thermal conductivity = 17.3 W/m·K) is 250°C. The 5 m diameter vessel has walls 2.54 cm thick. The internal and external heat-transfer coefficients are 723 W/m^2·K and 22.2 W/m^2·K, respectively. The ambient air temperature is 25°C.

35. What is the minimum insulation thickness (thermal conductivity = 0.057 W/m·K) that will reduce the external surface temperature of the insulation to 40°C?

(A) 2.54 cm
(B) 3.58 cm
(C) 3.90 cm
(D) 5.08 cm

36. The heat flux through the vessel wall after the insulation has been added is most nearly

(A) 236 W/m^2
(B) 333 W/m^2
(C) 36 700 W/m^2
(D) 153 000 W/m^2

Problems 37 and 38 are based on the following information.

4000 kg per hour of light lubricating oil are cooled by water in a shell-and-tube heat exchanger with oil on the tube side and water on the shell side. The oil temperature is reduced from 140°C to 50°C. Water is available at 20°C and a flow rate of 2.25 m^3/h. Data on the heat exchanger and materials are provided as follows.

$$\text{oil heat capacity} = 1670 \text{ J/kg·K}$$
$$\text{water heat capacity} = 4180 \text{ J/kg·K}$$
$$\text{water density} = 992 \text{ kg/m}^3$$
$$\text{overall heat-transfer coefficient} = 225 \text{ W/m}^2\text{·K}$$

37. The water exit temperature is most nearly

(A) 26.4°C
(B) 60.4°C
(C) 84.4°C
(D) 165°C

38. The heat-transfer surface area is most nearly

  (A) 9.05 m^2
  (B) 13.0 m^2
  (C) 14.9 m^2
  (D) 17.9 m^2

Problems 39 and 40 are based on the following information and illustration.

A self-enclosed hemisphere is made from a copper base (emissivity = 0.570) and a polished copper-nickel alloy dome (emissivity = 0.0590). The base is heated to maintain a constant 225°C, and the dome is at 115°C.

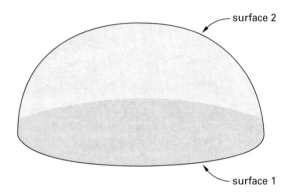

39. Determine the view factors F_{12}, F_{21}, and F_{22}.

  (A) $F_{12} = 1/3$, $F_{21} = 1/3$, and $F_{22} = 1/3$
  (B) $F_{12} = 1/\pi r^2$, $F_{21} = 1/\pi r$, and $F_{22} = 1/r$
  (C) $F_{12} = 1$, $F_{21} = 1/2$, and $F_{22} = 1/2$
  (D) $F_{12} = 1$, $F_{21} = 2$, and $F_{22} = 3/2$

40. Using the base as the reference, the radiation heat-transfer to the dome is most nearly

  (A) 13.9 W/m^2
  (B) 18.8 W/m^2
  (C) 206 W/m^2
  (D) 227 W/m^2

Problems 41 and 42 are based on the following information.

Aeration of a 500 m^3 wastewater pond is required for bio-remediation. Ten spargers located 4.60 m below the surface bubble compressed air through the pond at a rate of 0.425 m^3/min each. The mass transfer provided by each sparger is given as a transfer factor (the overall liquid mass-transfer coefficient multiplied by the ratio of interfacial area of bubbles to volume of solution) equal to 0.0680 h^{-1}. The Henry's law constant for dissolved oxygen is 4.06×10^9 Pa.

41. Calculate the equilibrium concentration of oxygen in the wastewater.

  (A) 9.29 mg/L
  (B) 11.3 mg/L
  (C) 21.6 mg/L
  (D) 44.2 mg/L

42. How long will it take to increase the dissolved oxygen from 3 mg/L to 6 mg/L?

  (A) 6.67 min
  (B) 39.6 min
  (C) 2.46 h
  (D) 6.60 h

43. Which of the following is not found on material safety data sheets?

  (A) chemical abstract series number
  (B) price
  (C) reactivity data
  (D) spill procedures

Problems 44–47 are based on the following information.

Gaseous ammonia is absorbed from an ammonia-air stream by water in a wetted-wall column. Pure water enters the top of the column where the ammonia partial pressure is 2000 Pa. The overall mass-transfer coefficient in terms of the partial pressure driving force is 9.20×10^{-3} mol/h·m^2·Pa. 70% of the total resistance to mass transfer is found in the gas phase. Equilibrium data for ammonia in water at operating conditions are provided as follows.

partial pressure NH$_3$ (Pa)	$\dfrac{\text{g of NH}_3}{100 \text{ g of water}}$
1533	1.20
2039	1.60
2572	2.00
3252	2.50
3945	3.00

44. The Henry's law constant for ammonia in water is most nearly

  (A) 1.17×10^5 Pa
  (B) 1.21×10^5 Pa
  (C) 1.24×10^5 Pa
  (D) 1.28×10^5 Pa

45. The individual gas-phase mass-transfer coefficient is most nearly

(A) 9.20×10^{-3} mol/h·m²·Pa
(B) 6.44×10^{-3} mol/h·m²·Pa
(C) 1.31×10^{-2} mol/h·m²·Pa
(D) 1.16×10^{-2} mol/h·m²·Pa

46. The individual liquid-phase mass-transfer coefficient is most nearly

(A) 233 mol/h·m²
(B) 2600 mol/h·m²
(C) 3620 mol/h·m²
(D) 3740 mol/h·m²

47. At the top of the column, the liquid ammonia mole fraction ($x_{\mathrm{NH_3}}$) and ammonia partial pressure ($p_{\mathrm{NH_3}}$) at the interface are most nearly

(A) $x_{\mathrm{NH_3}} = 0$ and $p_{\mathrm{NH_3}} = 2000$ Pa
(B) $x_{\mathrm{NH_3}} = 0.00508$ and $p_{\mathrm{NH_3}} = 595$ Pa
(C) $x_{\mathrm{NH_3}} = 0.0165$ and $p_{\mathrm{NH_3}} = 0$ Pa
(D) $x_{\mathrm{NH_3}} = 0.0790$ and $p_{\mathrm{NH_3}} = 857$ Pa

Problems 48 and 49 are based on the following information.

A mixture of hydrocarbons is separated in a single-stage flash tank held at 10°C and 620.5 kPa. The feed stream is a saturated liquid. The flow rates and equilibrium ratios of the various components are shown in the table.

component	substance	flow rate	equilibrium ratio
1	methane	10 mol/h	22.0
2	ethane	10 mol/h	3.60
3	n-octane	20 mol/h	0.00375

48. The flow rate of the exiting liquid stream is most nearly

(A) 16.8 mol/h
(B) 20.0 mol/h
(C) 23.2 mol/h
(D) 26.4 mol/h

49. The mole fraction composition of the exiting liquid stream (where x_i is the mole fraction of component i) is most nearly

(A) $x_1 = 0.0217$, $x_2 = 0.109$, and $x_3 = 0.996$
(B) $x_1 = 0.0255$, $x_2 = 0.120$, and $x_3 = 0.859$
(C) $x_1 = 0.250$, $x_2 = 0.250$, and $x_3 = 0.500$
(D) $x_1 = 0.561$, $x_2 = 0.431$, and $x_3 = 0.00322$

50. The following program was written to estimate the cosine of an angle accurate to six decimal places.

```
SET ERR = 10** - 6; N = 1; M = 1; SUM = 0
PI = 3.141593
   INPUT ANGLE
   X = ANGLE*PI/180
LOOPSTART
   W = X**(2N - 2)
   Z = (-1)**(N + 1)
   TERM = Z*W/M
   SUM = SUM + TERM
   IF ABS(TERM) < ERR THEN GO TO END
   N = N + 1
   M = M*(2N - 2)(2N - 3)
   GO TO LOOPSTART
END
   PRINT SUM, TERM
```

What will be the final value of N for an angle of 23° (entered as ANGLE)?

(A) 4
(B) 5
(C) 6
(D) 7

Problems 51 and 52 are based on the following information.

In the design of heat exchangers, it can be tedious to calculate the inside film coefficient if several scenarios are being evaluated. The equation for calculating this coefficient is

$$h_i = 0.023 \left(\frac{k}{D_i}\right) \left(\frac{D_i G}{\mu}\right)^{0.8} \left(\frac{C_p \mu}{k}\right)^{0.4}_b$$

An Excel spreadsheet used for calculating several values is represented in the following table.

	A	B
1	$k =$	0.59
2	$D_i =$	0.015
3	$G =$	1400
4	$\mu =$	0.0007
5	$C_p =$	4.02
6	$h_i =$	

51. For the given table, if generated using Excel, which equation should be input to calculate the results?

(A) (0.023*B1)/B2*((B2*B3)/B4)^0.8 *((B5*B4)*B1)^0.4
(B) (0.023*B1)/B2+((B2*B3)/B4)^0.8 +((B5*B4)/B1)^0.4
(C) (0.23*B1)/B2*((B2*B3)/B4)^0.8 *((B5*B4)/B1)^0.4
(D) (0.023*B1)/B2*((B2*B3)/B4)^0.8 *((B5*B4)/B1)^0.4

52. Assuming the correct units are used in the given table, the value of cell B6 will be most nearly

(A) 270
(B) 410
(C) 480
(D) 500

53. A laboratory technician is asked to prepare 15 L of a solution of potassium chloride (KCl) with a molarity of 0.2 mol/L. Approximately how much KCl will be required?

(A) 33 g
(B) 220 g
(C) 330 g
(D) 340 g

54. 2 mL of reagent A is used to titrate 20 mL of reagent B. What is the normality of reagent A if the normality of reagent B is 0.1 N?

(A) 0.1 N
(B) 0.5 N
(C) 1 N
(D) 2 N

55. 236 g of HCl are used to prepare a 10 L solution. The solution is at ambient conditions. The densities are 1.18 g/mL for HCl, 1 g/mL for H_2O, and 1.0036 g/mL for the solution. What is the most likely molality of this mixture?

(A) 0.60 mol/kg
(B) 0.66 mol/kg
(C) 0.70 mol/kg
(D) 0.76 mol/kg

56. What is the equilibrium constant of hydrofluoric acid, given the following equilibrium molarity concentrations?

$$HF = 0.2 \times 10^{-3} \text{ mol/L}$$
$$H^+ \text{ ion} = 1.0 \times 10^{-2} \text{ mol/L}$$
$$F^- \text{ ion} = 1.4 \times 10^{-5} \text{ mol/L}$$

(A) 2.8×10^{-10}
(B) 0.28×10^{-6}
(C) 7.0×10^{-4}
(D) 7.0×10^{-3}

57. 25 mL of a 0.01 mol/L sample of hydrochloric acid is slowly titrated with a 0.05 mol/L solution of sodium hydroxide. What volume of NaOH should be used?

(A) 0.50 ml
(B) 2.5 ml
(C) 5.0 ml
(D) 25 ml

58. In the given chemical equation, pure sodium has undergone what type of reaction?

$$2Na(s) + Cl_2(g) \rightarrow 2NaCl(s)$$

(A) reduction
(B) catalytic
(C) oxidation
(D) decomposition

Problems 59 and 60 are based on the following information.

10 kg/h of MeOH in 1000 kg/h of air are contacted with pure water at 15°C. The volumetric flow rate of the water stream is 50 m³/h. It is found that 95% of the incoming MeOH is removed by the water into the bottoms stream. 1% of the water goes overhead, and all of the air goes overhead.

59. The mass flow rate of the overhead stream is most nearly

(A) 1050 kg/h
(B) 1500 kg/h
(C) 1550 kg/h
(D) 1600 kg/h

60. The composition of MeOH in the bottoms stream is most nearly

(A) 0.020%
(B) 0.050%
(C) 1.9%
(D) 2.5%

SOLUTIONS

1. The rate expression for a first-order reaction is integrated to relate concentrations to time.

$$-\frac{dC_A}{dt} = kC_A$$
$$\int_{C_{A0}}^{C_A} \frac{dC_A}{C_A} = -\int_0^t kt$$
$$\ln \frac{C_A}{C_{A0}} = -kt$$

The rate constant is calculated from the half-life, the time when the concentration has been reduced to half of its original value.

$$\ln\left(\frac{0.5C_{A0}}{C_{A0}}\right) = -k(194\text{ s})$$

$$\ln 0.5 = -k(194\text{ s})$$

$$k = 3.57 \times 10^{-3}\text{ s}^{-1}$$

The value for k is substituted into the integrated rate equation to calculate the time of degradation. With 25% of the sample degraded,

$$C_A = (1 - 0.25)C_{A0} = 0.75C_{A0}$$

$$\ln\left(\frac{0.75C_{A0}}{C_{A0}}\right) = (-3.57 \times 10^{-3}\text{ s}^{-1})t$$

$$\ln 0.75 = (-3.57 \times 10^{-3}\text{ s}^{-1})t$$

$$t = 80.6\text{ s} \quad (81\text{ s})$$

The answer is A.

2. Determine the initial composition from the feed concentrations and component molecular weights.

$$\text{C}_2\text{H}_5\text{OH} + \text{CH}_3\text{COOH} \underset{k_2}{\overset{k_1}{\rightleftharpoons}} \text{CH}_3\text{COOC}_2\text{H}_5 + \text{H}_2\text{O}$$

$$\text{(A)} \qquad \text{(B)} \qquad\qquad \text{(R)} \qquad\qquad \text{(S)}$$

For the ethanol stream,

$$\left(800\ \frac{\text{g ethanol}}{\text{L}}\right)\left(\frac{1\text{ mol}}{46.1\text{ g}}\right) = 17.4\text{ mol/L ethanol}$$

$$\left(200\ \frac{\text{g water}}{\text{L}}\right)\left(\frac{1\text{ mol}}{18.0\text{ g}}\right) = 11.1\text{ mol/L water}$$

For the acetic acid (AA) stream,

$$\left(750\ \frac{\text{g AA}}{\text{L}}\right)\left(\frac{1\text{ mol}}{60.0\text{ g}}\right) = 12.5\text{ mol/L AA}$$

$$\left(250\ \frac{\text{g water}}{\text{L}}\right)\left(\frac{1\text{ mol}}{18.0\text{ g}}\right) = 13.9\text{ mol/L water}$$

The initial concentrations are

$$C_{A0} = \frac{\left(17.4\ \frac{\text{mol}}{\text{L}}\right)(1\text{ stream})}{2\text{ streams}} = 8.70\text{ mol/L}$$

$$C_{B0} = \frac{\left(12.5\ \frac{\text{mol}}{\text{L}}\right)(1\text{ stream})}{2\text{ streams}} = 6.25\text{ mol/L}$$

$$C_{R0} = 0\text{ mol/L}$$

$$C_{S0} = \frac{\left(11.1\ \frac{\text{mol}}{\text{L}}\right)(1\text{ stream}) + \left(13.9\ \frac{\text{mol}}{\text{L}}\right)(1\text{ stream})}{2\text{ streams}} = 12.5\text{ mol/L}$$

Acetic acid is the limiting reagent. The final concentrations are

$$C_B = C_{B0}(1 - x_B)$$

$$= \left(6.25\ \frac{\text{mol}}{\text{L}}\right)(1 - 0.35)$$

$$= 4.06\text{ mol/L}$$

$$C_A = C_{A0} - (C_{B0} - C_B)$$

$$= 8.70\ \frac{\text{mol}}{\text{L}} - \left(6.25\ \frac{\text{mol}}{\text{L}} - 4.06\ \frac{\text{mol}}{\text{L}}\right)$$

$$= 6.51\text{ mol/L}$$

$$C_R = C_{R0} + (C_{B0})(x_B)$$

$$= 0\ \frac{\text{mol}}{\text{L}} + \left(6.25\ \frac{\text{mol}}{\text{L}}\right)(0.35)$$

$$= 2.19\text{ mol/L}$$

$$C_S = C_{S0} + (C_{B0})(x_B)$$

$$= 12.5\ \frac{\text{mol}}{\text{L}} + \left(6.25\ \frac{\text{mol}}{\text{L}}\right)(0.35)$$

$$= 14.7\text{ mol/L}$$

The answer is B.

3. The feed stream flow rates are determined by calculating the space-time, τ. Space-time is the time required to process one reactor volume of feed.

$$\tau = \frac{V}{v_0} = \frac{C_{B0}V}{F_{B0}}$$

A material balance around the reactor yields

$$\text{input} - \text{output} = \text{disappearance by reaction}$$

$$F_{B0} - F_{B0}(1 - x_B) = -r_B V$$

$$F_{B0}x_B = -r_B V$$

For a second-order reversible reaction,

$$-r_B = k_1 C_A C_B - k_2 C_R C_S$$

Substituting gives

$$\tau = \frac{C_{B0}V}{F_{B0}} = \frac{C_{B0}X_B}{-r_B} = \frac{C_{B0} - C_B}{k_1 C_A C_B - k_2 C_R C_S}$$

$$\frac{V}{v_0} = \frac{C_{B0} - C_B}{k_1 C_A C_B - k_2 C_R C_S}$$

$$\frac{10\,000 \text{ L}}{v_0}$$

$$= \frac{6.25 \dfrac{\text{mol}}{\text{L}} - 4.06 \dfrac{\text{mol}}{\text{L}}}{\left(7.93 \times 10^{-6} \dfrac{\text{mol}}{\text{L}}\right)\left(6.51 \dfrac{\text{mol}}{\text{L}}\right)\left(4.06 \dfrac{\text{mol}}{\text{L}}\right)}$$
$$- \left(2.71 \times 10^{-6} \dfrac{\text{mol}}{\text{L}}\right)\left(2.19 \dfrac{\text{mol}}{\text{L}}\right)\left(14.7 \dfrac{\text{mol}}{\text{L}}\right)$$

$$v_0 = 0.559 \text{ L/s}$$

The flow rate for each stream is

$$\frac{v_0}{2} = \frac{0.559 \dfrac{\text{L}}{\text{s}}}{2} = 0.280 \text{ L/s} \quad (0.28 \text{ L/s})$$

The answer is A.

4. The activation energy, E_a, can be determined by rearranging the linearized Arrhenius expression.

$$\ln \frac{k_1}{k_2} = \left(-\frac{E_a}{R}\right)\left(\frac{1}{T_1} - \frac{1}{T_2}\right) = \left(\frac{E_a}{R}\right)\left(\frac{T_1 - T_2}{T_1 T_2}\right)$$

$$E_a = R \ln \frac{k_1}{k_2}\left(\frac{T_1 T_2}{T_1 - T_2}\right)$$

$$= \left(8.314 \frac{\text{J}}{\text{mol·K}}\right) \ln \left(\frac{0.0765 \dfrac{\text{L}}{\text{mol·s}}}{2.74 \dfrac{\text{L}}{\text{mol·s}}}\right)$$
$$\times \left(\frac{(623\text{K})(723\text{K})}{623\text{K} - 723\text{K}}\right)$$

$$= 134\,007 \text{ J/mol} \quad (134 \text{ kJ/mol})$$

The answer is D.

5. Assuming 1 mol total feed, the number of moles of each component at 99% conversion of NO_2 is calculated as in the table. The total number of moles does not change with reaction. Therefore, vol% composition equals mol% composition at all times.

component	initial moles	moles at 99% conversion	vol%
N_2	0.782	0.782	78.20
NO_2	0.084	$(0.01)(0.084) = 0.000\,84$	0.084
CO	0.134	$0.134 - 0.0832 = 0.0508$	5.08
NO	0.000	$(0.99)(0.084) = 0.0832$	8.32
CO_2	0.000	$(0.99)(0.084) = 0.0832$	8.32
total	1.000	1.0000	100.00

The answer is C.

6. The volume of 1 mol gas at 101 kPa and 400°C is required to calculate component concentration.

$$V = \frac{NRT}{p}$$

$$= \frac{(1 \text{ mol})\left(8.314 \dfrac{\text{L·kPa}}{\text{mol·K}}\right)(673\text{K})}{101 \text{ kPa}}$$

$$= 55.4 \text{ L}$$

The rate constant at 400°C can be calculated from a rearrangement of the Arrhenius expression.

$$\ln k_1 = \ln k_2 + \left(\frac{E_a}{R}\right)\left(\frac{1}{T_2} - \frac{1}{T_1}\right)$$

$$= \ln 2.74 + \left(\frac{134\,000 \dfrac{\text{J}}{\text{mol}}}{8.314 \dfrac{\text{J}}{\text{mol·K}}}\right)\left(\frac{1}{723\text{K}} - \frac{1}{673\text{K}}\right)$$

$$= -0.648$$

$$k_1 = e^{-0.648} = 0.523 \text{ L/mol·s}$$

The time required to reach 99% conversion is determined by solving the integrated second-order rate expression (designating NO_2 as A and CO as B).

$$-\frac{dC_A}{dt} = kC_A C_B = kC_A(C_{B0} - C_{A0} + C_A)$$

$$\int_0^t dt = \int_{C_{AO}}^{C_A} \left(\frac{dC_A}{kC_A(C_{B0} - C_{A0} + C_A)}\right)$$

$$= \frac{1}{k(C_{B0} - C_{A0})} \int_{C_{AO}}^{C_A}$$
$$\times \left(\frac{1}{C_A} - \frac{1}{C_{B0} - C_{A0} + C_A}\right) dC_A$$

$$t = \left(\frac{1}{k(C_{B0} - C_{A0})}\right)$$
$$\times \left(\ln \frac{C_A}{C_{A0}} - \ln \left(\frac{C_{B0} - C_{A0} + C_A}{C_{B0}}\right)\right)$$

$$= \left(\frac{1}{k(C_{B0} - C_{A0})}\right) \ln \frac{C_{A0} C_B}{C_A C_{B0}}$$

$$= \left(\frac{V}{k(n_{B0} - n_{A0})}\right) \ln \frac{n_{A0} n_B}{n_A n_{B0}}$$

$$= \left(\frac{55.4 \text{ L}}{\left(0.523 \dfrac{\text{L}}{\text{mol·s}}\right)(0.134 \text{ mol} - 0.084 \text{ mol})}\right)$$
$$\times \ln \left(\frac{(0.084 \text{ mol})(0.0508 \text{ mol})}{(0.00084 \text{ mol})(0.134 \text{ mol})}\right)$$

$$= 7701 \text{ s} \quad (128 \text{ min})$$

The answer is C.

7. The Hirschfelder equation relates diffusivity, pressure, and temperature.

$$D_{m_{T_2},p_2} = D_{m_{T_1},p_1}\left(\frac{p_1}{p_2}\right)\left(\frac{T_2}{T_1}\right)^{3/2}$$

$$= \left(1.98 \times 10^{-5}\ \frac{\text{m}^2}{\text{s}}\right)\left(\frac{101\ \text{kPa}}{150\ \text{kPa}}\right)\left(\frac{300\text{K}}{273\text{K}}\right)^{3/2}$$

$$= 1.54 \times 10^{-5}\ \text{m}^2/\text{s}$$

The answer is B.

8. The feed flow rate and composition must first be converted to a molar basis, where the mole fraction of water is considered.

$$F = \left(35\,380\ \frac{\text{kg}}{\text{h}}\right)\left(\frac{0.60\ \dfrac{\text{kg AA}}{\text{kg feed}}}{60.05\ \dfrac{\text{kg AA}}{\text{kmol}}} + \frac{0.40\ \dfrac{\text{kg water}}{\text{kg feed}}}{18.02\ \dfrac{\text{kg water}}{\text{kmol}}}\right)$$

$$= 1139\ \text{kmol/h}$$

$$x_F = \frac{\left(35\,380\ \dfrac{\text{kg feed}}{\text{h}}\right)\left(0.40\ \dfrac{\text{kg water}}{\text{kg feed}}\right)}{\left(18.02\ \dfrac{\text{kg water}}{\text{kmol}}\right)\left(1139\ \dfrac{\text{kmol}}{\text{h}}\right)}$$

$$= 0.6895$$

The bottoms (concentrated acetic acid) flow rate is calculated from a material balance around the column.

$$\text{moles: } F = D + B$$

$$\text{AA: } x_F F = x_D D + x_B B = x_D(F - B) + x_B B$$

$$B = \frac{F(x_D - x_F)}{x_D - x_B}$$

$$= \frac{\left(1139\ \dfrac{\text{kmol}}{\text{h}}\right)(0.90 - 0.6895)}{0.90 - 0.05}$$

$$= 282.1\ \text{kmol/h}$$

Converting to mass units,

$$\dot{m}_B = B(MW)$$

$$= \left(282.1\ \frac{\text{kmol}}{\text{h}}\right)$$

$$\times \left(\begin{array}{c}(0.95)\left(60.05\ \dfrac{\text{kg AA}}{\text{kmol}}\right)\\[2mm] + (0.05)\left(18.02\ \dfrac{\text{kg water}}{\text{kmol}}\right)\end{array}\right)$$

$$= 16\,347\ \text{kg/h}\quad(16\,300\ \text{kg/h})$$

The answer is D.

9. Reflux ratio is the ratio of reflux flow to overhead product flow. The minimum reflux ratio is determined graphically. Plot the equilibrium data provided along with the 45° line ($x = y$). The feed line for a saturated liquid is a vertical line at $y = x_F = 0.6895$. The minimum reflux ratio, $R_{\min}$, occurs when the rectifying operating line intersects the feed line at the equilibrium line. $R_{\min}$ is calculated from the slope of this line.

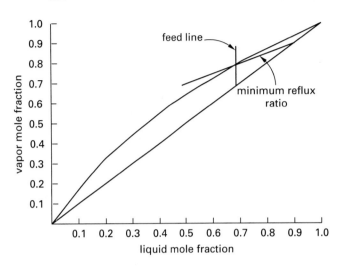

$$R_{\min} = \frac{\text{slope}}{1 - \text{slope}}$$

$$\text{slope} = \frac{y_1 - y_2}{x_1 - x_2} = \frac{0.900 - 0.790}{0.900 - 0.690} = 0.524$$

$$R_{\min} = \frac{0.524}{1 - 0.524} = 1.10$$

The answer is A.

10. Refer to the following illustration.

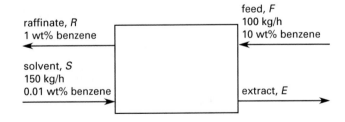

Given that the solvent and product are immiscible, a component mass balance yields

$$\text{cyclohexane: } (1 - x_F)F = (1 - x_R)R$$

$$R = \left(\frac{1 - x_F}{1 - x_R}\right)F$$

$$\text{furfural: } (1 - y_S)S = (1 - y_E)E$$

$$E = \left(\frac{1 - y_S}{1 - y_E}\right)S$$

The equilibrium expression $K = y_E/x_R$ may be incorporated.

$$E = \left(\frac{1 - y_S}{1 - Kx_R}\right) S$$

The total mass balance provides an expression that can be solved for the desired compositions.

$$F + S = R + E = \left(\frac{1 - x_F}{1 - x_R}\right) F + \left(\frac{Kx_R - y_S}{1 - Kx_R}\right) S$$

$$0 = \left(\frac{x_R - x_F}{1 - x_R}\right) F + \left(\frac{Kx_R - y_S}{1 - Kx_R}\right) S$$

Newton's method of root extraction is used to solve for x_R. First, an estimate is made for x_R. A new estimate is made based on

$$x_R^{j+1} = x_R^j - \frac{P(x)}{\left.\dfrac{\partial P(x)}{\partial x}\right|_{x=x_R^j}}$$

$$P(x) = \left(\frac{x_R - x_F}{1 - x_R}\right) F + \left(\frac{Kx_R - y_S}{1 - Kx_R}\right) S$$

$$= \left(\frac{x_R - 0.10}{1 - x_R}\right) \left(100 \; \frac{\text{kg}}{\text{h}}\right)$$

$$+ \left(\frac{0.680x_R - 0.0001}{1 - 0.680x_R}\right) \left(150 \; \frac{\text{kg}}{\text{h}}\right)$$

$$\frac{\partial P(x)}{\partial x_R} = \left(\frac{1 - x_F}{(1 - x_R)^2}\right) F + \left(\frac{K(1 - y_S)}{(1 - Kx_R)^2}\right) S$$

$$= \frac{90 \; \dfrac{\text{kg}}{\text{h}}}{(1 - x_R)^2} + \frac{102 \; \dfrac{\text{kg}}{\text{h}}}{(1 - 0.680x_R)^2}$$

Iterations are shown as follows.

x_R	$P(x_R)$	$\dfrac{\partial P(x)}{\partial x}$	$\dfrac{P(x)}{\dfrac{\partial P(x)}{\partial x}}$
0.0400	-20.7	205	-0.0101
0.0501	0.0217	209	1.05×10^{-4}
0.0500	8.17×10^{-4}	209	3.91×10^{-6}

$$x_R = 0.0500 \quad (5.00 \text{ wt\% benzene})$$
$$1 - x_R = 0.950 \quad (95.0 \text{ wt\% cyclohexane})$$

The answer is A.

11. The raffinate flow rate is calculated from the mass-balance equations derived in Prob. 10.

$$R = \left(\frac{1 - x_F}{1 - x_R}\right) F = \left(\frac{1 - 0.1}{1 - 0.05}\right) \left(100 \; \frac{\text{kg}}{\text{h}}\right)$$

$$= 94.7 \text{ kg/h}$$

The answer is A.

12. The number of theoretical stages and the extract composition are determined graphically on a right triangular ternary diagram (see illustration). The horizontal axis is the weight fraction of benzene in the product or solvent. The vertical axis is the weight fraction of extraction solvent. Because the solvent and product are immiscible, the horizontal axis is also the cyclohexane phase equilibrium line ($y = 0$ at all points). The furfural equilibrium line is obtained by mass balance. That is, the weight fraction sums with the benzene weight fraction to equal 1. The given solute concentrations are located on the illustration.

$$x_F = 0.10$$
$$x_R = 0.01$$
$$y_S = 0.0001$$

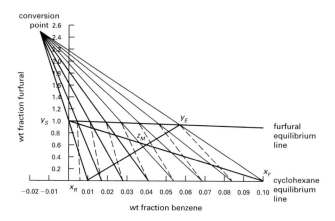

The mixing point weight fraction is calculated from a mass balance.

$$F + S = M$$
$$Fx_F + Sy_S = Mz_M$$

$$z_M = \frac{Fx_F + Sy_S}{F + S}$$

$$= \frac{\left(100 \; \dfrac{\text{kg}}{\text{h}}\right)(0.10) + \left(150 \; \dfrac{\text{kg}}{\text{h}}\right)(0.0001)}{100 \; \dfrac{\text{kg}}{\text{h}} + 150 \; \dfrac{\text{kg}}{\text{h}}}$$

$$= 0.0401$$

The mixing point falls on a straight line connecting x_F and y_S. The extract composition is determined by extending a straight line drawn through x_R and z_M to the furfural equilibrium line.

$$y_E = 0.057 = 5.7 \text{ wt\% benzene}$$
$$1 - y_E = 0.943 = 94.3 \text{ wt\% furfural}$$

The answer is C.

13. An Orsat analyzer does not directly measure water vapor content in the stack gas. The wet analysis of stack gas is calculated using nitrogen as a tie element. Any unaccounted-for oxygen is in the form of water vapor. Using a basis of 100 mol dry stack gas,

$$N_2 \text{ in air burned} = N_2 \text{ in stack gas}$$
$$= 80.4 \text{ mol}$$

$$N_{\text{air}} = \frac{N_{N_2}}{0.79 \ \frac{\text{mol } N_2}{\text{mol air}}} = \frac{80.4 \text{ mol } N_2}{0.79 \ \frac{\text{mol } N_2}{\text{mol air}}}$$
$$= 102 \text{ mol}$$

N_{O_2} in air burned

$$= N_{CO_2} + \frac{N_{CO}}{2} + N_{O_2} + \frac{N_{H_2O}}{2}$$

$$N_{H_2O} = 2 \left(\begin{array}{c} N_{O_2} \text{ air} \\ - \left(\begin{array}{c} N_{CO_2} \text{ stack} + \dfrac{N_{CO} \text{ flue}}{2} \\ + N_{O_2} \text{ stack} \end{array} \right) \end{array} \right)$$

$$= (2) \left(\begin{array}{c} (102 \text{ mol air}) \left(0.21 \ \dfrac{\text{mol } O_2}{\text{mol air}} \right) \\ - \left(10.6 \text{ mol } CO_2 + \dfrac{3.0}{2} \text{ mol } O \right) \\ + 6.0 \text{ mol } O_2 \end{array} \right)$$
$$= 6.6 \text{ mol } H_2O$$

The wet (total) stack gas composition is as follows.

component	number of moles (mol/100 mol dry stack gas)	mol % in wet stack gas
CO_2	10.6	9.9
CO	3.0	2.8
O_2	6.0	5.6
N_2	80.4	75.4
H_2O	6.6	6.2
total	106.6	

The amount of carbon burned is equal to the amount of carbon in the stack gas.

$$\frac{9.9 \text{ mol carbon} + 2.8 \text{ mol carbon}}{100 \text{ mol stack gas}}$$
$$= 12.7 \text{ mol carbon}/100 \text{ mol stack gas}$$

The answer is C.

14. The amount of air used on a wet (total) stack gas basis is

$$\frac{\text{mol}_{\text{air}}}{100 \text{ mol stack gas}} = \frac{\text{mol}_{N_2}}{0.79 \ \frac{\text{mol } N_2}{\text{mol air}}} = \frac{75.4 \text{ mol } N_2}{0.79 \ \frac{\text{mol } N_2}{\text{mol air}}}$$
$$= 95.4 \text{ mol air}/100 \text{ mol stack gas}$$

The answer is C.

15. The theoretical air requirement is that amount that would provide the stoichiometric amount of oxygen for complete combustion. Excess air is any amount over the theoretical requirement. The combustion equations are used to determine theoretical requirements.

$$C + O_2 \longrightarrow CO_2$$
$$H_2 + \tfrac{1}{2}O_2 \longrightarrow H_2O$$

12.7 mol carbon is burned (Prob. 14); therefore, 12.7 mol oxygen is required.

6.2 mol water is in the stack gas; therefore, 3.1 mol oxygen is required.

The theoretical air required is

$$\frac{\text{mol}_{O_2,\text{required}}}{0.21 \ \frac{\text{mol } O_2}{\text{mol air}}} = \frac{12.7 \text{ mol } O_2 + 3.1 \text{ mol } O_2}{0.21 \ \frac{\text{mol } O_2}{\text{mol air}}}$$
$$= 75.2 \text{ mol air required}$$

$$\% \text{ excess} = \left(\frac{\text{mol}_{\text{provided}} - \text{mol}_{\text{required}}}{\text{mol}_{\text{required}}} \right) \times 100\%$$
$$= \left(\frac{96.4 \text{ mol air} - 75.2 \text{ mol air}}{75.2 \text{ mol air}} \right) \times 100\%$$
$$= 26.9\% \quad (27\%)$$

The answer is D.

16. The number of moles of fuel entering the furnace is calculated from the ideal gas law.

$$N = \frac{pV}{RT} = \frac{(101\,300 \text{ Pa}) \left(45.0 \ \dfrac{\text{m}^3}{\text{d}} \right)}{\left(8.314 \ \dfrac{\text{m}^3 \cdot \text{Pa}}{\text{mol} \cdot \text{K}} \right) (25°C + 273°)}$$
$$= 1840 \text{ mol/d}$$

The number of moles of oxygen needed for complete combustion is calculated from the combustion equations as tabulated as follows.

$$CH_4(g) + 2 \ O_2(g) \longrightarrow CO_2(g) + 2 \ H_2O(l)$$
$$C_2H_6(g) + 3.5 \ O_2(g) \longrightarrow 2 \ CO_2(g) + 3 \ H_2O(l)$$
$$C_3H_8(g) + 5 \ O_2(g) \longrightarrow 3 \ CO_2(g) + 4 \ H_2O(l)$$
$$C_4H_{10}(g) + 6.5 \ O_2(g) \longrightarrow 4 \ CO_2(g) + 5 \ H_2O(l)$$

component	entrance mol%	entrance moles	moles O_2 required
methane	94.40	1737	3474
ethane	3.40	62.6	219
propane	0.60	11.0	55.0
butane	0.50	9.20	59.8
total			3808

Noting that air is 21% oxygen by volume and using 100% excess air,

$$\frac{mol_{air}}{d} = \frac{(2)\left(3808 \frac{mol\ O_2}{d}\right)}{0.21 \frac{mol\ O_2}{mol\ air}}$$

$$= 36\,267\ mol\ air/d$$

The volumetric flow rate is calculated using the ideal gas law.

$$V = \frac{NRT}{p}$$

$$= \frac{\left(36\,267 \frac{mol}{d}\right)\left(8.314 \frac{m^3 \cdot Pa}{mol \cdot K}\right)(298K)}{101\,300\ Pa}$$

$$= 887\ m^3/d$$

The answer is D.

17. The flue gas composition is determined from the entering components and combustion reactions as tabulated.

fuel	entrance (mol)	exit gas after combustion (mol)			
		CO_2	H_2O	O_2	N_2
methane	1737	1737	3474	–	–
ethane	62.6	125	188	–	–
propane	11.0	33.0	44.0	–	–
butane	9.20	36.8	46.0	–	–
carbon dioxide	11.0	11.0	–	–	–
nitrogen	9.20				9.20
air					
oxygen	7616	–	–	3808	–
nitrogen	28 650	–	–	–	28 650
total	38 110	1943	3752	3808	28 660

The total moles exiting in the flue gas is the sum of components.

$$total\ mol = mol\ CO_2 + mol\ H_2O + mol\ O_2 + mol\ N_2$$

$$= 1943\ mol + 3752\ mol + 3808\ mol$$

$$+ 28\,660\ mol$$

$$= 38\,163\ mol$$

The mol% composition can now be calculated.

$$mol\%\ CO_2 = \left(\frac{mol_{CO_2}}{N_{total}}\right) \times 100\%$$

$$= \left(\frac{1943\ mol}{38\,163\ mol}\right) \times 100\%$$

$$= 5.09\ mol\%$$

$$mol\%\ H_2O = \left(\frac{mol_{H_2O}}{N_{total}}\right) \times 100\%$$

$$= \left(\frac{3752\ mol}{38\,163\ mol}\right) \times 100\%$$

$$= 9.83\ mol\%$$

$$mol\%\ O_2 = \left(\frac{mol_{O_2}}{N_{total}}\right) \times 100\%$$

$$= \left(\frac{3808\ mol}{38\,163\ mol}\right) \times 100\%$$

$$= 9.98\ mol\%$$

$$mol\%\ N_2 = \left(\frac{mol_{N_2}}{N_{total}}\right) \times 100\%$$

$$= \left(\frac{28\,660\ mol}{38\,163\ mol}\right) \times 100\%$$

$$= 75.1\ mol\%$$

The answer is A.

18. The maximum wall temperature is achieved under adiabatic conditions, when no heat escapes to the surroundings.

$$\Delta H = \sum_F n_F \Delta H_C + \sum_{out} n_i h_i(T_{ad}) - \sum_{in} n_i h_i(T_{feed})$$

$$= 0$$

Burning the richest possible fuel (that with the highest heating value) with no excess air will create the most heat, thus yielding the highest temperature. A mass balance must be completed as in Probs. 16 and 17.

component	mol%	amount (mol)	O_2 required (mol)	products (mol) CO_2	products (mol) H_2O	$n_F \Delta H_c$ heat of combustion (kJ)
methane	85.5	1573	3146	1573	3146	-1.401×10^6
ethane	10.0	184	644	368	552	-2.870×10^5
propane	3.0	55.2	276	166	221	-1.225×10^5
butane	1.0	18.4	120	73.6	92.0	-5.288×10^4
pentane	0.5	9.2	73.6	46.0	55.2	-3.250×10^4
total	100	1840	4260	2227	4066	-1.896×10^6

Taking the reference temperature as 25°C,

$$\sum_{in} n_i h_i(25°C) = 0$$

The outlet enthalpy can be calculated from the heat of evaporation of water found in the steam tables and the component specific heat, integrated from the reference temperature (25°C) to the adiabatic temperature (T_{ad}).

$$\sum_{\text{out}} n_i h_i = n_{H_2O}(\Delta H_{\text{vap}})_{H_2O} \text{ at } 25°C$$

$$+ \int_{25°C}^{T_{ad}} \left(\sum n_i c_{pi}\right) dT$$

$$= (4066 \text{ mol})\left(2442 \frac{J}{g}\right)\left(18.0 \frac{g}{\text{mol}}\right) + \int_{25°C}^{T_{ad}}$$

$$\times \begin{pmatrix} (2227 \text{ mol } CO_2) \\ \times \left(36.80 \frac{J}{\text{mol·°C}} + (1.406 \times 10^{-2})T\right) \\ + (4066 \text{ mol } H_2O) \\ \times \left(33.43 \frac{J}{\text{mol·°C}} + (8.392 \times 10^{-3})T\right) \\ + (4260 \text{ mol } O_2) \\ \times \left(\frac{0.79 \text{ mol } N_2}{0.21 \text{ mol } O_2}\right) \\ \times \left(28.99 \frac{J}{\text{mol·°C}} + (2.729 \times 10^{-3})T\right) \end{pmatrix} dT$$

$$= 1.787 \times 10^8 \text{ J}$$

$$+ \int_{25°C}^{T_{ad}} \left(6.825 \times 10^5 \frac{J}{\text{mol·°C}} + 109.2T\right) dT$$

$$= 1.616 \times 10^8 \text{ J} + \left(6.825 \times 10^5 \frac{J}{\text{mol·°C}}\right) T_{ad}$$

$$+ \left(54.60 \frac{J}{\text{mol·°C}}\right) T_{ad}^2$$

The enthalpy equation becomes
$$\Delta H = 0$$

$$= -1.734 \times 10^6 \text{ kJ} + \left(682.5 \frac{kJ}{°C}\right) T_{ad}$$

$$+ \left(0.054\,60 \frac{kJ}{°C^2}\right) T_{ad}^2$$

$$T_{ad} = \frac{-682.5 \frac{kJ}{°C} + \sqrt{\left(682.5 \frac{kJ}{°C}\right)^2 - (4)\left(0.054\,60 \frac{kJ}{°C^2}\right) \times (-1.735 \times 10^6 \text{ kJ})}}{(2)\left(0.054\,60 \frac{kJ}{°C^2}\right)}$$

$$= 2167°C \quad (2170°C)$$

The answer is B.

19. The flow rate for the exiting flue gas is determined by summing the total moles tabulated in Prob. 18 and using the ideal gas law.

$$N_{\text{total}} = \text{mol}_{CO_2} + \text{mol}_{H_2O} + \text{mol}_{N_2}$$
$$= 2227 \text{ mol } CO_2 + 4066 \text{ mol } H_2O$$
$$+ (4260 \text{ mol } O_2)\left(\frac{0.79 \text{ mol } N_2}{0.21 \text{ mol } O_2}\right)$$
$$= 22\,319 \text{ mol}$$

$$V = \frac{NRT}{p}$$
$$= \frac{(22\,319 \text{ mol})\left(8.314 \frac{m^3·Pa}{\text{mol·K}}\right)(2439K)}{101\,300 \text{ Pa}}$$
$$= 4468 \text{ m}^3/\text{d} \quad (4470 \text{ m}^3/\text{d})$$

The answer is B.

20. The value of an item after depreciation is its book value.
$$BV = C - \sum D_j$$

Using the accelerated cost recovery system, depreciation in any year j is calculated as follows.
$$D_j = (\text{factor})_j C$$

C is the initial cost, neglecting any scrap value. The factors depend on the cost recovery period. The book value after 3 yr, using a cost recovery period of 10 yr, is

$$BV_3 = C - \sum_1^3 C(\text{factor})_j$$
$$= \$27,500 - (\$27,500)(0.10 + 0.18 + 0.144)$$
$$= \$15,840 \quad (\$15,800)$$

The answer is A.

21. The present value of the investment is calculated from the costs and incomes over the life of the process (15 yr).

$$\begin{aligned}\text{present value} &= -\text{investment} \\ &+ (\text{income} - \text{expenses})(P/A, 12\%, 15) \\ &- (\text{upgrade})\left[\begin{matrix}(P/F, 12\%, 5) \\ + (P/F, 12\%, 10)\end{matrix}\right] \\ &+ (\text{salvage})(P/F, 12\%, 15) \\ &= -\$189,000 + (\$38,000 - \$3520)(6.8109) \\ &- (\$7500)(0.5674 + 0.3220) \\ &+ (\$63,000)(0.1827) \\ &= \$50,679 \quad (\$50,700)\end{aligned}$$

The answer is B.

22. Inflation deflates the value of the dollar. Inflation can be accounted for by shifting the interest used in the economic factors according to the following formula.

$$d = i + f + if$$
$$= 0.12 + 0.03 + (0.12)(0.03)$$
$$= 0.1536$$

No table is available for an interest rate of 15.36%, so the factors are calculated.

$$(P/A, 15.36\%, 15) = \frac{(i+1)^n - 1}{i(1+i)^n}$$
$$= \frac{(1.1536)^{15} - 1}{(0.1536)(1.1536)^{15}}$$
$$= 5.7470$$
$$(P/F, 15.36\%, 5) = (1+i)^{-n}$$
$$= (1.1536)^{-5}$$
$$= 0.4895$$
$$(P/F, 15.36\%, 10) = (1+i)^{-n}$$
$$= (1.1536)^{-10}$$
$$= 0.2396$$
$$(P/F, 15.36\%, 15) = (1+i)^{-n}$$
$$= (1.1536)^{-15}$$
$$= 0.1173$$

Using these factors, the present value is calculated.

$$PV = -\$189,000 + (\$34,480)(5.7470)$$
$$- (\$7500)(0.4895 + 0.2396)$$
$$+ (\$63,000)(0.1173)$$
$$= \$11,078 \quad (\$11,100)$$

The answer is A.

23. Each alternative has an equal lifetime (10 yr), so they can be compared using the present value and an interest rate of 8%.

$$P_A = A(P/A, 8\%, 10)$$
$$= (\$2000)(6.7101)$$
$$= \$13,420$$
$$P_B = P$$
$$= \$14,500$$
$$P_C = P + F(P/F, 8\%, 5)$$
$$= \$7000 + (\$10,000)(0.6806)$$
$$= \$13,806$$
$$P_D = P + F(P/F, 8\%, 5)$$
$$= \$10,000 + (\$7000)(0.6806)$$
$$= \$14,764$$

Alternative A (the current method) has the lowest present value, so no investment should be made.

The answer is A.

24. The annual investments grow to an amount that is then distributed over the next 30 yr. The retirement annuity is

$$A_R = A_I(F/A, 8\%, 40)(A/P, 8\%, 30)$$
$$= (\$2000)(259.0565)(0.0888)$$
$$= \$46,008 \quad (\$46,000)$$

The answer is D.

25. Because the alternatives have unequal lives, they are compared by the annual cost method (capital recovery method). The initial investment (P) for each unit is redistributed over its lifetime (n yr) using the capital recovery factor.

The annuity is then summed with the annual maintenance cost (AMC) to obtain an equivalent uniform annual cost (EUAC).

$$EUAC = A + AMC$$
$$= P(A/P, i\%, n) + AMC$$
$$EUAC(A) = (\$20,000)(0.1770) + \$1800$$
$$= \$5340$$
$$EUAC(B) = (\$25,000)(0.1770) + \$1200$$
$$= \$5625$$
$$EUAC(C) = (\$30,000)(0.1468) + \$1200$$
$$= \$5604$$
$$EUAC(D) = (\$45,000)(0.1339) + \$500$$
$$= \$6526$$

Alternative A is the most economical.

The answer is A.

26. The closed-loop transfer function is

$$y = \left(\frac{G_p G_c}{1 + G_p G_c}\right) r$$
$$= \left(\frac{4s\left(1 + \frac{1}{s^n}\right)}{1 + 4s\left(1 + \frac{1}{s^n}\right)}\right) r$$
$$\frac{y}{r} = \frac{4s + 4s^{1-n}}{1 + 4s + 4s^{1-n}}$$

The behavior of a transfer function at steady state is calculated by taking the limit as s approaches 0. For steady-state setpoint tracking, y/r must equal 1 as s approaches 0.

For $n = 0$,

$$\lim_{s\to 0}\left(\frac{y}{r}\right) = \frac{4s+4s}{1+4s+4s} = 0$$

For $n = 1$,

$$\lim_{s\to 0}\left(\frac{y}{r}\right) = \frac{4s+4}{1+4s+4} = \frac{4}{5}$$

For $n = 2$,

$$\lim_{s\to 0}\left(\frac{y}{r}\right) = \lim_{s\to 0}\left(\frac{4s+4s^{-1}}{1+4s+4s^{-1}}\right)$$

Apply l'Hôpital's rule.

$$\lim_{s\to 0}\left(\frac{y}{r}\right) = \lim_{s\to 0}\left(\frac{\left(\frac{d}{ds}\right)(4s+4s^{-1})}{\left(\frac{d}{ds}\right)(1+4s+4s^{-1})}\right)$$

$$= \lim_{s\to 0}\left(\frac{4-4s^{-2}}{4-4s^{-2}}\right)$$

$$= 1$$

For $n = 3$,

$$\lim_{s\to 0}\left(\frac{y}{r}\right) = \lim_{s\to 0}\left(\frac{4s+4s^{-2}}{1+4s+4s^{-2}}\right)$$

Applying l'Hôpital's rule,

$$\lim_{s\to 0}\left(\frac{y}{r}\right) = \lim_{s\to 0}\left(\frac{\left(\frac{d}{ds}\right)(4s+4s^{-2})}{\left(\frac{d}{ds}\right)(1+4s+4s^{-2})}\right)$$

$$= \lim_{s\to 0}\left(\frac{4-8s^{-3}}{4-8s^{-3}}\right)$$

$$= 1$$

Therefore, the minimum value of n for offset-free setpoint tracking by the output is 2.

The answer is C.

27. The closed-loop transfer function relating y and r is

$$y = \left(\frac{G_pG_c}{1+G_pG_c}\right)r = \left(\frac{\left(\frac{1}{2s+1}\right)\left(\frac{K}{s+2}\right)}{1+\left(\frac{1}{2s+1}\right)\left(\frac{K}{s+2}\right)}\right)r$$

$$\frac{y}{r} = \frac{K}{2s^2+5s+K+2}$$

A second-order transfer function in standard form is expressed as follows.

$$\frac{kw^2}{s^2+2w\zeta s+w^2}$$

For a critically damped second-order response, the damping factor, ζ, equals 1.

$$\frac{kw^2}{s^2+2w\zeta s+w^2} = \frac{K}{2s^2+5s+K+2}$$

$$= \frac{\frac{1}{2}K}{s^2+\frac{5}{2}s+\frac{K+2}{2}}$$

Equating terms in the denominator,

$$w = \sqrt{\frac{K+2}{2}}$$

$$2w\zeta = \frac{5}{2}$$

$$\zeta = \frac{5}{4w}$$

$$= \frac{5}{4}\sqrt{\frac{2}{K+2}}$$

$$= 1$$

$$\left(\frac{25}{16}\right)\left(\frac{2}{K+2}\right) = 1$$

$$K+2 = \frac{50}{16}$$

$$K = \frac{18}{16}$$

$$= 1.125 \quad (1.13)$$

The answer is B.

28. A first-order plus dead time transfer function has the following form.

$$G_p = \frac{K_pe^{-t_ds}}{\tau s+1}$$

The process gain, K_p, can be measured as the ratio of the change in output to the change in input at steady state.

$$K_p = \frac{\Delta \text{output}_{ss}}{\Delta \text{input}_{ss}}$$
$$= \frac{5-0}{3-1}$$
$$= 2.5$$

The dead time, t_d, is the time between the movement of the input and the first reaction of the process. From the problem illustration, $t_d = 2$. The process time constant, τ, is the time constant for the first-order lag exhibited in the output response. One time constant after the output starts moving, the output will have traveled $(1 - e^{-1}) \times 100\% = 63\%$ from the first steady state to the second. From the graph, this is approximately one time unit. Therefore, the process model is most nearly

$$G_p = \frac{2.5e^{-2s}}{s+1}$$

The answer is A.

29. The shear stress at the wall is calculated from the following expression.

$$\tau = \left(\frac{D}{4}\right)\left(-\frac{\Delta p}{L}\right)$$
$$= \left(\frac{0.0525 \text{ m}}{4}\right)\left(\frac{2.08 \times 10^5 \text{ Pa}}{300 \text{ m}}\right)$$
$$= 9.10 \text{ Pa}$$

The answer is D.

30. Viscosity is calculated by equating two expressions for shear stress.

$$\tau = -\mu \frac{dv}{dr} = \left(\frac{r}{2}\right)\left(-\frac{\Delta p}{L}\right)$$
$$-\int_{v_c}^{v} dv = \left(-\frac{\Delta p}{2\mu L}\right)\int_0^r r\, dr$$
$$v_c - v = \frac{\Delta p r^2}{4\mu L}$$

This shear stress formula is only valid for laminar flow. This can be checked with the Reynolds number.

$$\text{Re} = \frac{\rho v D}{\mu}$$
$$= \frac{\left(800 \; \frac{\text{kg}}{\text{m}^3}\right)\left(0.702 \; \frac{\text{m}}{\text{s}}\right)(0.0525 \text{ m})}{0.105 \text{ Pa·s}}$$
$$= 302$$

Since the Reynolds number is less than 2100, the flow is in the laminar regime for flow in a pipe and the formula for shear stress is valid.

The no-slip condition at the wall implies $v = 0$ at $r = R$ and allows calculation of v_c.

$$v_c = -\frac{\Delta p R^2}{4\mu L}$$
$$v = \left(-\frac{\Delta p}{4\mu L}\right)(R^2 - r^2)$$

The average velocity is given in the problem statement.

$$v_{avg} = \frac{Q}{A} = \frac{\int v\, dA}{\int dA}$$
$$= \frac{\int_0^R v(2\pi r)dr}{\int_0^R (2\pi r)dr}$$

Substituting for v and v_c,

$$v_{avg} = \frac{Q}{\pi R^2} = \frac{(-\Delta p)2\pi}{4\mu L \pi R^2}\int_0^R (R^2 - r^2)r\, dr$$
$$= \left(\frac{-\Delta p}{2\mu L R^2}\right)\left(\frac{R^4}{4}\right)$$
$$= \left(\frac{-\Delta p}{8\mu L}\right)R^2$$

Rearranging to solve for viscosity,

$$\mu = \left(\frac{-\Delta p}{8L}\right)R^2\left(\frac{\pi R^2}{Q}\right) = \frac{-\Delta p \pi D^4}{128 L Q}$$
$$= \frac{(2.08 \times 10^5 \text{ Pa})\pi(0.0525 \text{ m})^4}{(128)(300 \text{ m})\left(4.44 \; \frac{\text{m}^3}{\text{h}}\right)\left(\frac{1 \text{ h}}{3600 \text{ s}}\right)}$$
$$= 0.105 \text{ Pa·s}$$

The answer is A.

31. The fluid velocity 1 cm from the pipe wall is calculated from the equation derived in Prob. 30.

$$v = \left(-\frac{\Delta p}{4\mu L}\right)(R^2 - r^2) = \left(-\frac{\Delta p}{16\mu L}\right)(D^2 - d^2)$$
$$= \left(\frac{2.08 \times 10^5 \text{ Pa}}{(16)(0.105 \text{ Pa·s})(300 \text{ m})}\right)$$
$$\times \left((0.0525 \text{ m})^2 - (0.0325 \text{ m})^2\right)$$
$$= 0.702 \text{ m/s}$$

The answer is B.

32. The power equation is as follows.

$$P = \frac{\dot{m}gh_f}{\eta}$$

Rearranging to solve for the mass flow rate,

$$\dot{m} = \frac{P\eta}{gh_f}$$

$$= \frac{(13\,000 \text{ W})(0.65)\left(60\ \frac{\text{s}}{\text{min}}\right)}{\left(9.81\ \frac{\text{m}}{\text{s}^2}\right)(25\text{ m})}$$

$$= 2067 \text{ kg/min}$$

Under ambient conditions, the specific gravity, S, is 1.0 and the density, ρ, is 1000 kg/m³. Therefore, the volumetric flow rate is

$$Q = \dot{m}\rho$$

$$= \left(2067\ \frac{\text{kg}}{\text{min}}\right)\left(\frac{1\text{ m}^3}{1000\text{ kg}}\right)\left(60\ \frac{\text{min}}{\text{h}}\right)$$

$$= 124 \text{ m}^3/\text{h}\quad(120\text{ m}^3/\text{h})$$

The answer is B.

33. Velocity is determined from the following equation.

$$\text{v} = \frac{Q}{A}$$

Therefore, the pipe area can be determined by rearranging.

$$A = \frac{Q}{\text{v}}$$

The velocity is known to be 0.6 m/s. The desired volumetric flow rate should be converted from m³/h to m³/s.

$$Q = \left(250\ \frac{\text{m}^3}{\text{h}}\right)\left(\frac{1\text{ h}}{3600\text{ s}}\right)$$

$$= 0.0694 \text{ m}^3/\text{s}$$

Substituting,

$$A = \frac{0.0694\ \dfrac{\text{m}^3}{\text{s}}}{0.6\ \dfrac{\text{m}}{\text{s}}}$$

$$= 0.116 \text{ m}^2$$

The pipe area is derived from the pipe diameter as follows.

$$A = \frac{\pi D^2}{4}$$

Rearranging to solve for the pipe diameter,

$$D = \sqrt{\frac{4A}{\pi}} = \sqrt{\frac{(4)(0.116\text{ m}^2)}{\pi}}$$

$$= 0.384 \text{ m}\quad(380\text{ mm})$$

The answer is B.

34. The mach number is determined from

$$M = \frac{\text{v}}{c}$$

Calculate the velocity of sound from

$$c = \sqrt{kRT}$$

First, determine the specific gas constant.

$$R = \frac{\overline{R}}{\text{MW}} = \frac{8314\ \dfrac{\text{J}}{\text{kmol·K}}}{28\ \dfrac{\text{g}}{\text{mol}}}$$

$$= 0.297 \text{ J/g·K}$$

$$T = 200°\text{C} + 273° = 473\text{K}$$

Substituting,

$$c = \sqrt{(1.28)\left(0.297\ \frac{\text{J}}{\text{g·K}}\right)(473\text{K})}$$

$$= 334 \text{ m/s}$$

Now rearrange the mach number equation to solve for velocity.

$$\text{v} = Mc$$

$$= (0.1)\left(334\ \frac{\text{m}}{\text{s}}\right)$$

$$= 33.4 \text{ m/s}\quad(33\text{ m/s})$$

The answer is C.

35. The curvature of the vessel wall is small enough that it can be neglected.

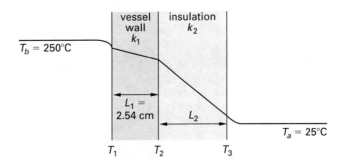

The heat flux through any segment of the composite is the same as the heat flux through the combination.

$$\frac{q}{A} = h_i(T_b - T_1) = \left(\frac{k_1}{L_1}\right)(T_1 - T_2)$$

$$= \left(\frac{k_2}{L_2}\right)(T_2 - T_3) = h_o(T_3 - T_a)$$

$$= \frac{T_b - T_a}{\dfrac{1}{h_i} + \dfrac{L_1}{k_1} + \dfrac{L_2}{k_2} + \dfrac{1}{h_o}}$$

$$h_o(T_3 - T_a) = \frac{T_b - T_a}{\dfrac{1}{h_i} + \dfrac{L_1}{k_1} + \dfrac{L_2}{k_2} + \dfrac{1}{h_o}}$$

$$\left(22.2 \,\frac{\text{W}}{\text{m}^2\text{·K}}\right)(40°\text{C} - 25°\text{C})$$

$$= \frac{(250°\text{C} - 25°\text{C})}{\dfrac{1}{723 \,\frac{\text{W}}{\text{m}^2\text{·K}}} + \dfrac{0.0254 \text{ m}}{17.3 \,\frac{\text{W}}{\text{m}^2\text{·K}}}}$$

$$+ \frac{L_2}{0.057 \,\frac{\text{W}}{\text{m}^2\text{·K}}} + \frac{1}{22.2 \,\frac{\text{W}}{\text{m}^2\text{·K}}}$$

$$L_2 = 0.0358 \text{ m} \quad (3.58 \text{ cm})$$

The answer is B.

36. The heat flux through the tank wall equals the heat flux from the external surface to the ambient air.

$$\frac{q}{A} = \left(\frac{k_1}{L_1}\right)(T_1 - T_2) = h_o(T_3 - T_2)$$

$$= \left(22.2 \,\frac{\text{W}}{\text{m}^2\text{·K}}\right)(40°\text{C} - 25°\text{C})$$

$$= 333 \text{ W/m}^2$$

The answer is B.

37. The heat flux into the shell side and out the tube side is

$$\dot{Q} = \dot{m}c_p \Delta T$$

On the oil (tube) side,

$$\dot{Q} = \left(4000 \,\frac{\text{kg}}{\text{h}}\right)\left(\frac{1 \text{ h}}{3600 \text{ s}}\right)\left(1670 \,\frac{\text{J}}{\text{kg·K}}\right)$$

$$\times (140°\text{C} - 50°\text{C})$$

$$= 167\,000 \text{ W}$$

On the water (shell) side,

$$\dot{Q} = \dot{m}c_p(T_{so} - T_{si})$$

$$T_{so} = \frac{q}{\dot{m}c_p} + T_{si}$$

$$= \frac{167\,000 \text{ W}}{\left(2.25 \,\frac{\text{m}^3}{\text{h}}\right)\left(992 \,\frac{\text{kg}}{\text{m}^3}\right)\left(\frac{1 \text{ h}}{3600 \text{ s}}\right)\left(4180 \,\frac{\text{J}}{\text{kg·K}}\right)}$$

$$+ 20°\text{C}$$

$$= 84.4°\text{C}$$

The answer is C.

38. Heat flux for the heat exchanger is expressed as follows.

$$q = UA\Delta T_{\text{lm}}$$

Rearranging and using the subscripts s for the shell fluid and t for tube fluid,

$$A = \frac{q}{U\Delta T_{\text{lm}}}$$

$$= \left(\frac{q}{U}\right)\left(\frac{\ln \dfrac{T_{to} - T_{si}}{T_{ti} - T_{so}}}{(T_{to} - T_{si}) - (T_{ti} - T_{so})}\right)$$

$$= \left(\frac{167\,000 \text{ W}}{225 \,\frac{\text{W}}{\text{m}^2\text{·K}}}\right)\left(\frac{\ln \dfrac{50.0°\text{C} - 20.0°\text{C}}{140°\text{C} - 84.4°\text{C}}}{\begin{array}{c}(50.0°\text{C} - 20.0°\text{C}) \\ - (140°\text{C} - 84.4°\text{C})\end{array}}\right)$$

$$= 17.9 \text{ m}^2$$

The answer is D.

39. The base faces only the dome, so $F_{12} = 1$.

The reciprocity relationship states

$$A_1 F_{12} = A_2 F_{21}$$

$$F_{21} = \left(\frac{A_1}{A_2}\right)F_{12} = \left(\frac{\pi r^2}{2\pi r^2}\right) \quad (1)$$

$$= 1/2$$

The summation rule stipulates that

$$F_{21} + F_{22} = 1$$

$$F_{22} = 1 - F_{21} = 1 - \frac{1}{2}$$

$$= 1/2$$

The answer is C.

40. The relationship for radiation heat-transfer between two gray surfaces that form an enclosure is

$$\dot{Q}_{12} = \frac{\sigma(T_1^4 - T_2^4)}{\dfrac{1 - \epsilon_1}{\epsilon_1 A_1} + \dfrac{1}{A_1 F_{12}} + \dfrac{1 - \epsilon_2}{\epsilon_2 A_2}}$$

$$\frac{\dot{Q}_{12}}{A_1} = \frac{\sigma(T_1^4 - T_2^4)}{\dfrac{1 - \epsilon_1}{\epsilon_1} + \dfrac{1}{F_{12}} + \dfrac{1 - \epsilon_2}{\epsilon_2 \dfrac{A_2}{A_1}}}$$

$$= \frac{\left(5.68 \times 10^{-8} \ \dfrac{\text{W}}{\text{m}^2 \cdot \text{K}^4}\right)\left((498\text{K})^4 - (388\text{K})^4\right)}{\dfrac{1 - 0.57}{0.57} + \dfrac{1}{1} + \dfrac{1 - 0.059}{(0.059)\left(\dfrac{2\pi r^2}{\pi r^2}\right)}}$$

$$= 227 \ \text{W/m}^2$$

The answer is D.

41. The average hydrostatic pressure of the rising air bubbles is taken as the mean of the pressure at the top of the pond and the pressure at sparger level.

$$p_{\text{top}} = 1.013 \times 10^5 \ \text{Pa}$$

$$\begin{aligned} p_{\text{sparger}} &= p_{\text{top}} + \rho g h \\ &= 1.013 \times 10^5 \ \text{Pa} \\ &\quad + \left(1000 \ \frac{\text{kg}}{\text{m}^3}\right)\left(9.81 \ \frac{\text{m}}{\text{s}^2}\right)(4.6 \ \text{m}) \\ &= 1.464 \times 10^5 \ \text{Pa} \end{aligned}$$

$$\begin{aligned} p_{\text{mean}} &= \frac{1.013 \times 10^5 \ \text{Pa} + 1.464 \times 10^5 \ \text{Pa}}{2} \\ &= 1.24 \times 10^5 \ \text{Pa} \end{aligned}$$

Next, the partial pressure of oxygen in the bubbles is calculated.

$$\begin{aligned} p_{O_2} &= p y_{O_2} = (1.24 \times 10^5 \ \text{Pa})(0.21) \\ &= 2.60 \times 10^4 \ \text{Pa} \end{aligned}$$

The equilibrium concentration of oxygen in solution is related to the partial pressure by Henry's law.

$$p_{O_2} = h x_{O_2}$$

$$\begin{aligned} x_{O_2} &= \frac{p_{O_2}}{h} = \frac{2.60 \times 10^4 \ \text{Pa}}{4.06 \times 10^9 \ \text{Pa}} \\ &= 6.40 \times 10^{-6} \end{aligned}$$

The equilibrium mole fraction can be converted to mg/L.

$$C_{O_2}^* = x_{O_2} C_{\text{solution}}(\text{MW})_{O_2}$$

$$= (6.40 \times 10^{-6})\left(\frac{997 \ \dfrac{\text{g H}_2\text{O}}{\text{L}}}{18.0 \ \dfrac{\text{g H}_2\text{O}}{\text{mol}}}\right)\left(32.0 \ \frac{\text{g O}_2}{\text{mol}}\right)$$

$$= 0.0113 \ \text{g/L} \quad (11.3 \ \text{mg/L})$$

The answer is B.

42. Mass transfer can be expressed in terms of the overall liquid mass-transfer coefficient.

$$\frac{N_{O_2}}{A} = K_L(C_{O_2}^* - C_{O_2,L})$$

Expressed in differential form, the equation becomes

$$\frac{dN_{O_2}}{dt} = K_L A(C_{O_2}^* - C_{O_2,L})$$

In a constant-volume pond, the equation can be expressed in terms of concentration.

$$\frac{dC_{O_2,L}}{dt} = K_L \frac{A}{V}(C_{O_2}^* - C_{O_2,L})$$

Separating variables and integrating yields

$$\int_{C_{O_2,L|0}}^{C_{O_2,L|t}} \frac{dC_{O_2,L}}{C_{O_2}^* - C_{O_2,L}} = K_L \frac{A}{V} \int_0^t dt$$

$$\ln \frac{C_{O_2}^* - C_{O_2,L|0}}{C_{O_2}^* - C_{O_2,L|t}} = K_L \frac{A}{V} t$$

$$\begin{aligned} t &= \ln \frac{C_{O_2}^* - C_{O_2,L|0}}{C_{O_2}^* - C_{O_2,L|t}}\left(\frac{1}{K_L \dfrac{A}{V}}\right) \\ &= \ln \frac{11.3 \ \dfrac{\text{mg}}{\text{L}} - 3.0 \ \dfrac{\text{mg}}{\text{L}}}{11.3 \ \dfrac{\text{mg}}{\text{L}} - 6.0 \ \dfrac{\text{mg}}{\text{L}}} \\ &\quad \times \left(\frac{1}{\left(\dfrac{0.0680}{\text{h}\cdot\text{sparger}}\right)(10 \ \text{spargers})}\right)\left(60 \ \frac{\text{min}}{\text{h}}\right) \\ &= 39.6 \ \text{min} \end{aligned}$$

The answer is B.

43. Price is not included on material safety data sheets.

The answer is B.

44. Henry's law, valid at low concentrations as x approaches 0, states that the partial pressure of a gas is proportional to the concentration in a liquid at equilibrium.

$$p_i^* = hx_i^*$$

To determine the Henry's law constant, calculate the mole fraction of NH_3.

$$x_{NH_3} = \cfrac{Y\left(\cfrac{1\ g\ NH_3}{100\ g\ H_2O}\right)\left(\cfrac{1\ mol\ NH_3}{17\ g\ NH_3}\right)}{Y\left(\cfrac{1\ g\ NH_3}{100\ g\ H_2O}\right)\left(\cfrac{1\ mol\ NH_3}{17\ g\ NH_3}\right) + \left(\cfrac{100\ g\ H_2O}{100\ g\ H_2O}\right)\left(\cfrac{1\ mol\ H_2O}{18\ g\ H_2O}\right)}$$

Next, calculate the proportionality constant, $h = p/x$, for the data provided and extrapolate to $x = 0$.

p_{NH_3} (Pa)	Y	x_{NH_3}	h (Pa)
1533	1.20	0.0126	121 700
2039	1.60	0.0167	122 100
2572	2.00	0.0207	124 300
3252	2.50	0.0258	126 000
3945	3.00	0.0308	128 100

Plotting h versus x and extrapolating to 0 as in the following illustration, Henry's law constant is 117 000.

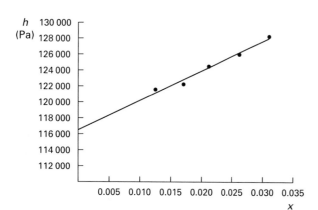

The answer is A.

45. The overall mass-transfer coefficient incorporates the total resistance to mass transfer in both phases.

$$\text{total resistance} = \frac{1}{K_G}$$

$$= \frac{1}{9.20 \times 10^{-3}\ \dfrac{mol}{h \cdot m^2 \cdot Pa}}$$

$$= 109\ h \cdot m^2 \cdot Pa/mol$$

The resistance in the gas phase, $1/k_G$, is 70% of the total resistance.

$$\frac{1}{k_G} = 0.70 \frac{1}{K_G}$$

$$= (0.70)\left(\frac{1}{9.20 \times 10^{-3}\ \dfrac{mol}{h \cdot m^2 \cdot Pa}}\right)$$

$$= 76.1\ h \cdot m^2 \cdot Pa/mol$$

$$k_G = 1.31 \times 10^{-2}\ mol/h \cdot m^2 \cdot Pa$$

The answer is C.

46. The liquid-phase mass-transfer coefficient can be calculated from the gas phase, overall mass-transfer coefficients, and Henry's law constant as follows.

$$\frac{1}{K_G} = \frac{1}{k_G} + \frac{h}{k_L}$$

$$k_L = \left(\left(\frac{1}{K_G} - \frac{1}{k_G}\right)\left(\frac{1}{h}\right)\right)^{-1}$$

$$= \left(\left(\frac{1}{9.20 \times 10^{-3}\ \dfrac{mol}{h \cdot m^2 \cdot Pa}} - \frac{1}{1.31 \times 10^{-2}\ \dfrac{mol}{h \cdot m^2 \cdot Pa}}\right)\left(\frac{1}{1.17 \times 10^5\ Pa}\right)\right)^{-1}$$

$$= 3616\ mol/h \cdot m^2 \quad (3620\ mol/h \cdot m^2)$$

The answer is C.

47. The mass flux across the interface can be determined from the overall mass-transfer coefficient.

$$N_{NH_3} = K_G(p_{NH_3,bulk} - p^*_{NH_3})$$

The partial pressure of ammonia in the bulk is given as 2000 Pa. The equilibrium partial pressure is calculated using Henry's law.

$$p^*_{NH_3} = hx$$

$$= (11.17 \times 10^5\ Pa)(0)$$

$$= 0$$

Thus, the flux is

$$N_{NH_3} = K_G p_{NH_3}$$

$$= \left(9.20 \times 10^{-3}\ \frac{mol}{h \cdot m^2 \cdot Pa}\right)(2000\ Pa)$$

$$= 18.4\ mol/h \cdot m^2$$

The interface concentration is determined from the individual mass-transfer coefficients.

The gas phase is

$$N_{NH_3} = k_G(p_{NH_3,bulk} - p_{NH_3,i})$$

$$p_{NH_3,i} = p_{NH_3,bulk} - \frac{N_{NH_3}}{k_G}$$

$$= 2000 \text{ Pa} - \frac{18.4 \, \frac{\text{mol}}{\text{h·m}^2}}{0.0131 \, \frac{\text{mol}}{\text{h·m}^2\text{·Pa}}}$$

$$= 595 \text{ Pa}$$

The liquid phase is

$$N_{NH_3} = k_L(x_{NH_3,i} - x_{NH_3,bulk})$$

$$x_{NH_3,i} = x_{NH_3,bulk} + \frac{N_{NH_3}}{k_L}$$

$$= 0 + \frac{18.4 \, \frac{\text{mol}}{\text{h·m}^2}}{3620 \, \frac{\text{mol}}{\text{h·m}^2}}$$

$$= 0.005\,08$$

The answer is B.

Solutions 48 and 49 are based on the following information.

Material balance and vapor-liquid equilibrium calculations are used to determine exit flow rates and compositions.

$$F = V + L$$

$$f_i = v_i + l_i \quad [i = 1,2,3]$$

$$Fz_i = Vy_i + Lx_i \quad [i = 1,2,3]$$

$$y_i = K_i x_i = \frac{v_i}{V} \quad [i = 1,2,3]$$

$$x_i = \frac{l_i}{L} = \frac{f_i - v_i}{L} \quad [i = 1,2,3]$$

The iterative calculations are simplified by taking a basis of 1.0 mol/h feed ($F = 1.0$ mol/h). Rearranging the equations gives an expression for the vapor fraction as a function of vapor flow rate.

$$y_i = \frac{z_i K_i}{1 + V(K_i - 1)}$$

Calculation proceeds as follows: First assume a value for V, then calculate the vapor fraction of each component. The sum of the vapor fractions should equal 1.00, otherwise, assume another value for V. After a few attempts, the value for V can be extrapolated.

Assume $V = 0.5$ mol/h.

component	z_i	K_i	y_i
1	0.25	22.0	0.478
2	0.25	3.60	0.391
3	0.50	0.003 75	0.003 74
total	1.00		0.873

Assume $V = 0.4$ mol/h.

component	z_i	K_i	y_i
1	0.25	22.0	0.585
2	0.25	3.60	0.441
3	0.50	0.003 75	0.003 12
total	1.00		1.029

Interpolate between estimates to obtain vapor flow rate where the sum of vapor fractions equal 1.00.

$$\frac{1.029 - 0.873}{0.4 \, \frac{\text{mol}}{\text{h}} - 0.5 \, \frac{\text{mol}}{\text{h}}} = \frac{1.029 - 1.00}{0.4 \, \frac{\text{mol}}{\text{h}} - V}$$

$$V = 0.419 \text{ mol/h} \quad \text{[at a basis of 1 mol/h feed]}$$

The actual feed rate is

$$F = 10 \, \frac{\text{mol}}{\text{h}} + 10 \, \frac{\text{mol}}{\text{h}} + 20 \, \frac{\text{mol}}{\text{h}}$$

$$= 40 \text{ mol/h}$$

Calculate the actual vapor and liquid exit rates from the material balances.

$$V = \left(0.419 \, \frac{\text{mol}}{\text{h}}\right)\left(\frac{40 \, \frac{\text{mol}}{\text{h}}}{1 \, \frac{\text{mol}}{\text{h}}}\right)$$

$$= 16.8 \text{ mol/h}$$

$$L = F - V = 40 \, \frac{\text{mol}}{\text{h}} - 16.8 \, \frac{\text{mol}}{\text{h}}$$

$$= 23.2 \text{ mol/h}$$

Calculate the liquid composition from equilibrium expressions.

Assume $V = 0.419$ mol/h.

component	z_i	K_i	y_i	$x_i = \dfrac{y_i}{K_i}$
1	0.25	22.0	0.561	0.0255
2	0.25	3.60	0.431	0.120
3	0.50	0.003 75	0.003 22	0.859
total	1.00		0.995	1.005

48. The answer is C.

49. The answer is B.

50. The values for the variables at each iteration are as follows.

iteration	N	M	W	Z	TERM	SUM
1	1	1	1	1	1	1
2	2	2	0.161143	-1	-8.06×10^{-2}	0.919429
3	3	24	0.025967	1	1.08×10^{-3}	0.920511
4	4	720	0.004184	-1	-5.81×10^{-6}	0.920505
5	5	40,320	0.000674	1	1.67×10^{-8}	0.920505

When $N = 5$ the absolute value of TERM is less than 10^{-6}.

The answer is B.

51. The Excel input equation shown in option D will calculate the results properly.

The answer is D.

52. The value of cell B6 can be determined from

$$h_i = 0.023 \frac{k}{D_i} \left(\frac{D_i G}{\mu} \right)^{0.8} \left(\frac{C_p \mu}{k} \right)^{0.4}_b$$

$$= (0.023) \left(\frac{0.59}{0.015} \right) \left(\frac{(0.015)(1400)}{0.0007} \right)^{0.8}$$

$$\times \left(\frac{(4.02)(0.0007)}{0.59} \right)^{0.4}_b$$

$$= 407 \quad (410)$$

The answer is B.

53. Molarity is determined from the number of moles of solute per volume of solution in liters. So in this case,

$$M = \frac{N_{KCl}}{V_{solution}}$$

Rearranging and solving for the number of moles,

$$N_{KCl} = \left(0.20 \ \frac{mol}{L} \right) (15 \ L)$$

$$= 3 \ mol$$

Converting this value from moles to grams,

$$KCl = (3.0 \ mol) \left(74.6 \ \frac{g}{mol} \right)$$

$$= 223.8 \ g \quad (220 \ g)$$

The answer is B.

54. By definition, the product of the normality and the volume of reagent A equals the product of the normality and the volume of reagent B.

$$N_A V_A = N_B V_B$$

Convert the 2 mL volume of reagent A and the 20 mL volume of reagent B to liters, and solve for the normality of reagent A.

$$N_A = \frac{N_B V_B}{V_A}$$

$$= \frac{\left(0.1 \ \frac{eq}{L} \right) (0.02 \ L)}{0.002 \ L}$$

$$= 1 \ eq/L \quad (1 \ N)$$

The answer is C.

55. Molality, m, is determined from the number of moles of solute, N, per mass of solvent, m, in kilograms. So in this case,

$$m = \frac{N_{HCl}}{m_{H_2O}}$$

Since the amount of HCl is given in grams, convert this value to moles.

$$N_{HCl} = (236 \ g) \left(\frac{1 \ mol}{36.5 \ g} \right)$$

$$= 6.47 \ mol$$

Determine the mass of the solution in grams. First convert the 10 L volume of solution to milliliters, and then obtain the mass of the solution using the 1.0036 g/mL density given.

$$m_{solution} = V \rho$$

$$= (10\,000 \ mL) \left(1.0036 \ \frac{g}{mL} \right)$$

$$= 10\,036 \ g$$

To obtain the mass of solvent, subtract the mass of HCl from the mass of solution.

$$m_{H_2O} = m_{solution} - m_{HCl}$$

$$= 10\,036 \ g - 236 \ g$$

$$= 9800 \ g$$

Convert the 9800 g mass of solution to kilograms, and solve for molality.

$$m = \frac{6.47 \ mol}{9.8 \ kg}$$

$$= 0.66 \ mol/kg$$

The answer is B.

56. When placed in a diluent such as water, HF dissociates into H^+ ions and F^- ions. Therefore, at equilibrium, the equilibrium constant can be expressed as

$$K_{eq} = \frac{[H^+][F^-]}{[HF]}$$

$$= \frac{\left[1 \times 10^{-2} \frac{mol}{L}\right]\left[1.4 \times 10^{-5} \frac{mol}{L}\right]}{0.2 \times 10^{-3} \frac{mol}{L}}$$

$$= 7 \times 10^{-4}$$

The answer is C.

57. Use the following chemical equations.

$$HCl \rightarrow H^+ + Cl^-$$
$$NaOH \rightarrow Na^+ + OH^-$$

The number of moles of OH^- equals the number of moles of H^+.

Use the following equation to obtain the number of moles of H^+.

$$N_{H^+} = M_{HCl}V_{HCl}$$
$$= \left(0.01 \frac{mol}{L}\right)(0.025 \, L)$$
$$= 2.5 \times 10^{-4} \, mol$$

Use the following equation to obtain the number of moles of OH^-.

$$N_{OH^-} = M_{NaOH}V_{NaOH}$$

Rearranging to solve for the volume of NaOH,

$$V_{NaOH} = \frac{2.5 \times 10^{-4} \, mol}{0.05 \frac{mol}{L}}$$
$$= 0.005 \, L \quad (5 \, mL)$$

The answer is C.

58. In the given chemical equation, pure sodium has gone from a neutral, or 0, state of charge to a positive state of charge. Assign oxidation numbers as follows.

$$\overset{0}{2Na} + \overset{0}{Cl_2} \rightarrow \overset{+1 \; -1}{2Na\,Cl}$$

The answer is C.

59. Perform mass balances of each component on an hourly basis.

Air:
$$\dot{m}_i = 1000 \, kg/h$$
$$\dot{m}_B = 0 \, kg/h$$
$$\dot{m}_o = 1000 \, kg/h$$

MeOH:
$$\dot{m}_i = 10 \, kg/h$$
$$\dot{m}_B = \dot{m}_i 0.95 = \left(10 \frac{kg}{h}\right)(0.95)$$
$$= 9.5 \, kg/h$$
$$\dot{m}_o = \dot{m}_i - \dot{m}_B = 10 \frac{kg}{h} - 9.5 \frac{kg}{h}$$
$$= 0.5 \, kg/h$$

Water:

Convert $50 \, m^3/h$ to kg/h. At 15°C, the specific gravity is 1.0. The density is $1000 \, kg/m^3$.

$$\dot{m}_i = Q\rho = \left(50 \frac{m^3}{h}\right)\left(1000 \frac{kg}{m^3}\right)$$
$$= 50\,000 \, kg/h$$
$$\dot{m}_o = \dot{m}_i 0.01 = \left(50\,000 \frac{kg}{h}\right)(0.01)$$
$$= 500 \, kg/h$$
$$\dot{m}_B = \dot{m}_i - \dot{m}_o = 50\,000 \frac{kg}{h} - 500 \frac{kg}{h}$$
$$= 49\,500 \, kg/h$$

The sum of all components in the overhead is given in the following table.

component	overhead (kg/h)
air	1000
MeOH	0.50
water	500
total	1500.5

The answer is B.

60. Perform mass balances of each component on an hourly basis, concentrating on the bottoms stream.

Air:
$$\dot{m}_B = 0 \, kg/h$$

MeOH:

95% of the incoming MeOH goes into the bottoms stream:

$$\dot{m}_B = \dot{m}_i 0.95$$

$$\dot{m}_i = 10 \ \frac{kg}{h}$$

$$\dot{m}_B = \left(10 \ \frac{kg}{h}\right)(0.95)$$

$$= 9.5 \ kg/h$$

Water:

1% of the incoming water goes into the overhead stream; therefore 99% of the incoming water goes into the bottoms stream.

$$\dot{m}_i = Q\rho$$

$$= \left(50 \ \frac{m^3}{h}\right)\left(1000 \ \frac{kg}{m^3}\right)$$

$$= 50\,000 \ kg/h$$

$$\dot{m}_B = \dot{m}_i 0.99$$

$$= \left(50\,000 \ \frac{kg}{h}\right)(0.99)$$

$$= 49\,500 \ kg/h$$

To obtain the total flow rate of the bottoms stream, sum all of the components.

component	bottoms (kg/h)
air	0.0
MeOH	9.5
water	49 500.0
total	49 509.5

Therefore the composition is

$$composition = \left(\frac{9.5 \ \frac{kg}{h}}{49\,509.5 \ \frac{kg}{h}}\right) \times 100\%$$

$$= 0.02\%$$

The answer is A.

Practice Exam 2

PROBLEMS

Problems 1–5 are based on the following information.

In the first-order, isothermal, aqueous-phase reaction $A \to B$, the half-life of A is 1000 s. A conversion of 90% is desired to minimize waste. It is desired to process 1000 kg/h of A. The molecular weight of A is 100 g/mol and the molar density of A is 1 mol/L.

1. The rate constant for the reaction is most nearly

(A) $0.00001 \ m^3/mol \cdot s$
(B) $0.0002 \ s^{-1}$
(C) $0.0007 \ s^{-1}$
(D) $0.002 \ mol/m^3 \cdot s$

2. If the reactor is operated as a batch reactor and it is initially charged with pure A, the reactor volume required, neglecting charge and discharge time, is most nearly

(A) $5.0 \ m^3$
(B) $10 \ m^3$
(C) $40 \ m^3$
(D) $190 \ m^3$

3. If the reactor is operated as a continuous stirred tank reactor (CSTR) with a pure A feed, the required volume for the same conditions would be most nearly

(A) $5.0 \ m^3$
(B) $10 \ m^3$
(C) $40 \ m^3$
(D) $190 \ m^3$

4. If the batch reactor has a charge time of 3 h and a discharge time of 3 h, which mode of operation of the reactor will have the smallest volume?

(A) batch
(B) CSTR
(C) both
(D) neither

5. If instead of one CSTR, five equal-sized CSTRs were to be used under the same conditions, their total volume would be most nearly

(A) $3.0 \ m^3$
(B) $6.0 \ m^3$
(C) $12 \ m^3$
(D) $36 \ m^3$

6. In the isothermal first-order gas-phase reaction $A \to B + C$, the equation to be used for a PFR is

(A) $k\tau = \int_0^{X_A} (1 + \varepsilon_A X_A) \, dX_A$

(B) $k\tau = -\ln(1 - X_A)$

(C) $k\tau = \int_0^{X_A} \dfrac{(1 + \varepsilon_A X_A)^2 \, dX_A}{1 - X_A}$

(D) $k\tau = \int_0^{X_A} \dfrac{(1 + \varepsilon_A X_A) \, dX_A}{1 - X_A}$

Problems 7 and 8 are based on the following information.

The oxidation of sulfur dioxide proceeds at 538°C and 5 bar (500 kPa) in a bomb calorimeter.

$$SO_2 + \frac{1}{2}O_2 \rightleftharpoons SO_3$$

Data for the components is

component	state	heat of formation at 25°C (cal/mol)	free energy of formation at 25°C (cal/mol)
SO_2	gas	−70 940	−71 680
O_2	gas	0	0
SO_3	gas	−94 390	−88 590

7. The boss needs an estimate for a meeting today, so she will allow the assumption that the enthalpy of formation is constant at any temperature. With this assumption the free-energy change of the reaction at 538°C is most nearly

(A) −5700 cal/mol
(B) −1300 cal/mol
(C) +1300 cal/mol
(D) +5700 cal/mol

8. Assuming a stoichiometric initial ratio of sulfur dioxide and oxygen, and no sulfur trioxide, the equilibrium conversion to sulfur trioxide at 538°C and 5 bar (500 kPa) is most nearly

(A) 55%
(B) 64%
(C) 82%
(D) 93%

Problems 9 and 10 are based on the following information.

The Redlich-Kwong equation of state may be written as [10].

$$V = \frac{RT}{P} + b - \frac{a(V-b)}{T^{1/2}PV(V+b)}$$

$$a = \frac{0.42748R^2T_c^{2.5}}{p_c}$$

$$b = \frac{0.08664RT_c}{p_c}$$

The critical properties of dichloromethane are

$$T_c = 510.0K$$

$$p_c = 60.8 \text{ bar}$$

A vessel containing dichloromethane is at equilibrium with the temperature and pressure being

$$T = 60°C$$

$$P = 10 \text{ bar}$$

9. Using the ideal gas law as a starting guess, the vapor molar volume, V, of dichloromethane in the vessel is most nearly

(A) 550 cm³/mol
(B) 1300 cm³/mol
(C) 2100 cm³/mol
(D) 2800 cm³/mol

10. For the liquid molar volume, a more convenient form of the equation is

$$V = \frac{V^3 - \frac{RT}{P}V^2 - \frac{ab}{PT^{1/2}}}{c}$$

$$c = b^2 + \frac{bRT}{P} - \frac{a}{PT^{1/2}}$$

Using b as a starting guess, the liquid molar volume, V, of dichloromethane in the vessel is most nearly

(A) 60 cm³/mol
(B) 75 cm³/mol
(C) 80 cm³/mol
(D) 550 cm³/mol

11. A pseudocode program is written to solve two equations in two unknowns, with X and Y in units of radians.

```
SET X = 3.
DO J = 1 TO 50
    Y = (2. - 2. * SIN(X) ) / 3.
    XC = (3. - 2. * COS(Y) ) / 3.
    ERROR = (XC - X) / X
    OUTPUT J, X, Y, ERROR
    IF (ABS(error) < 0.1 ) THEN STOP
    X = XC
NEXT J
STOP
```

The output for J, X, Y, and ERROR, respectively, when this program is finished is most nearly

(A) 1   3.00   0.573   −0.853
(B) 2   0.440   0.383   −0.132
(C) 3   0.382   0.418   0.0241
(D) 4   0.391   0.413   −0.00394

Problems 12–15 are based on the following information.

It is desired to heat cold ethanol in a double-pipe heat exchanger using hot water. Data are as follows.

parameter	cold ethanol inside pipe	hot water outside pipe
inner diameter (cm)	5.250	7.793
outer diameter (cm)	6.0325	8.89
inlet temperature (°C)	20	80
outlet temperature (°C)	40	50
flow rate (kg/s)	1.26	–
heat capacity (J/kg·K)	2500	4200
thermal conductivity (W/m·K)	0.182	0.689
viscosity (cp)	0.25	0.12
fouling factor (m²·K/W)	0	0.000 35
heat-transfer coefficient (W/m²·K)	–	3200

The double-pipe heat exchanger operated in counter-current flow mode is illustrated as follows.

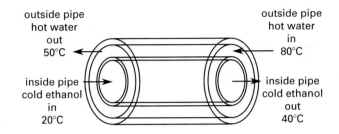

12. The hot water flow is most nearly

(A) 0.50 kg/s
(B) 1.5 kg/s
(C) 2.5 kg/s
(D) 3.5 kg/s

13. The best flow configuration and log mean temperature difference (LMTD) is most nearly

(A) cocurrent, 28°C
(B) countercurrent, 35°C
(C) cocurrent, 35°C
(D) countercurrent, 28°C

14. The inside heat-transfer coefficient is most nearly

(A) 10 W/m²·K
(B) 850 W/m²·K
(C) 970 W/m²·K
(D) 1400 W/m²·K

15. Neglecting inner tube wall heat transfer, the overall heat-transfer coefficient and exchanger area based on the outside area of the inner tube, assuming the best flow configuration is used, is most nearly

(A) 1020 W/m²·K and 1.1 m²
(B) 680 W/m²·K and 0.50 m²
(C) 680 W/m²·K and 2.7 m²
(D) 1020 W/m²·K and 15 m²

Problems 16–19 are based on the following information.

It is desired to distill an acetone-water stream in a plate column with a total condenser and a partial reboiler. The feed consists of 1000 mol/s of 50 mol% acetone and is at a temperature such that its quality is 50% liquid and 50% vapor. 98% of the acetone in the feed stream is to be recovered. The overhead product (distillate) composition is to be 90 mol% acetone. The operating external reflux ratio is to be eight times the minimum external reflux ratio. Constant molal overflow is assumed.

Equilibrium data for the acetone-water system is given below [8].

acetone in liquid (mole fraction)	acetone in vapor (mole fraction)
0	0
0.05	0.6381
0.10	0.7301
0.15	0.7716
0.20	0.7916
0.30	0.8124
0.40	0.8269
0.50	0.8387
0.60	0.8532
0.70	0.8712
0.80	0.8950
0.90	0.9335
0.95	0.9627
1	1

The vapor-liquid equilibrium data is as shown.

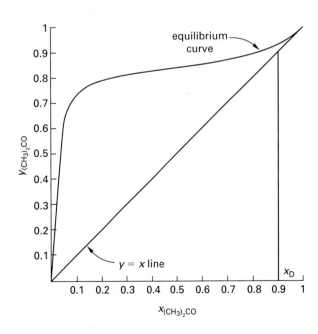

The column piping arrangement is shown.

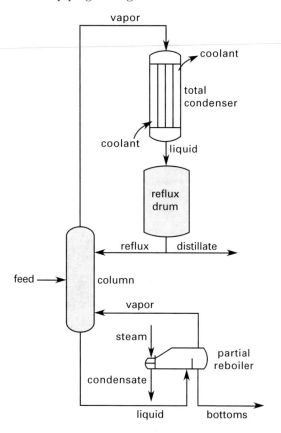

16. The bottoms acetone composition is most nearly

    (A) 0.10 mol%
    (B) 1.6 mol%
    (C) 2.2 mol%
    (D) 4.3 mol%

17. The minimum external reflux ratio is most nearly

    (A) 0.18
    (B) 0.70
    (C) 1.5
    (D) 3.9

18. The liquid-to-vapor molar flow ratio in the stripping section is most nearly

    (A) 1.1
    (B) 1.5
    (C) 2.1
    (D) 3.7

19. Assuming a Murphree plate efficiency of 70%, the number of stages is most nearly

    (A) three stages plus a partial reboiler
    (B) five stages plus a partial reboiler
    (C) eight stages plus a partial reboiler
    (D) nine stages plus a partial reboiler

Problems 20–24 are based on the following information.

In the liquid-phase process, toluene, cobalt catalyst, and air are fed into a hot pressurized reactor, which reacts to form benzoic acid [2]. The reaction mixture goes to a distillation column, which sends recycled, unreacted toluene overhead back to the reactor. Water is removed from the condensed reactor hot off-gas stream. Distillate bottoms benzoic acid is sent for hot water washing, precipitation, and filtering. The reaction is

$$C_6H_5CH_3 + \frac{3}{2}O_2 \rightarrow C_6H_5COOH + H_2O$$

Conversion of toluene in the reactor is 40% per pass. The overall process yield is 90% based on toluene. The product stream from the column bottoms product contains 1000.0 kg/h benzoic acid and some toluene.

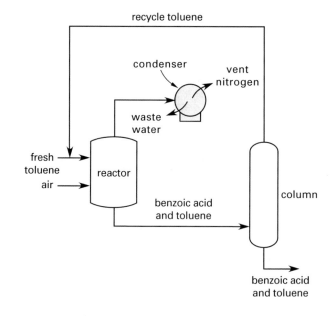

20. The benzoic acid and toluene product flow rate is most nearly

    (A) 1000 kg/h
    (B) 1100 kg/h
    (C) 1300 kg/h
    (D) 1400 kg/h

21. The wastewater flow rate is most nearly

    (A) 150 kg H_2O/h
    (B) 170 kg H_2O/h
    (C) 250 kg H_2O/h
    (D) 350 kg H_2O/h

22. The theoretical air flow rate is most nearly

(A) 530 kg air/h
(B) 1700 kg air/h
(C) 1900 kg air/h
(D) 2500 kg air/h

23. The theoretical air at STP conditions, 0°C and 1 atm (101.33 kPa), volumetric flow rate is most nearly

(A) 100 m³/h
(B) 300 m³/h
(C) 590 m³/h
(D) 1300 m³/h

24. The recycle toluene flow rate is most nearly

(A) 1000 kg/h
(B) 2300 kg/h
(C) 3200 kg/h
(D) 3500 kg/h

Problems 25–27 are based on the following information.

A waste feed stream (1) from the plant of volumetric flow rate $F_1 = 10000$ L/d is to be treated in an activated sludge reactor as illustrated.

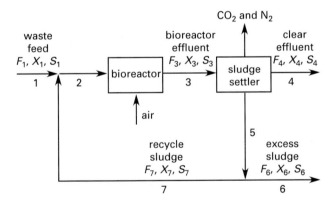

This waste feed stream (2) is combined with recycle sludge (7) to form the feed stream (2) to an aerated bioreactor. If the culture is aerobic, the overall reaction is

substrate + oxygen → biomass + carbon dioxide + water

After the bioreactor, a settler clarifies the sludge into a clear effluent (4) discharged to the environment and a settled sludge (5). The settled sludge (5) is bled of excess sludge (6) with the remaining sludge recycled (7) to the reactor. The waste feed (1) to the system has a substrate concentration in terms of the 5 d biological oxygen demand (BOD_5) of 500 mg/L, and the bioreactor outlet (3) has been specified by OSHA to

be 10 mg/L. Tests show that the biomass sludge concentration returning in the recycle (7) is 1.3 times the biomass concentration of the flow leaving the reactor (3). The effluent substrate (4) and recycled substrate (5) have the same concentration as the reactor effluent (3). Tests were performed on this bioreactor and data on the kinetics are as follows. The rate of biomass production is

$$r_X \left[\frac{\text{mg biomass}}{\text{d·L}} \right] = \left(\frac{\mu_{\max} S}{K_S + S} - k_e \right) X$$

S substrate concentration   mg substrate/L
X biomass concentration   mg biomass/L

The substrate consumption rate is

$$r_S \left[\frac{\text{mg substrate}}{\text{d·L}} \right] = \left(\frac{\mu_{\max} S}{K_S + S} \right) \left(\frac{X}{Y_{X/S}} \right)$$

The kinetic parameters for the waste were determined as

$\mu_{\max}$	maximum specific growth rate	2.5 d⁻¹
K_S	Monod coefficient	50 mg substrate/L
k_e	cell maintenance rate	0.05 d⁻¹
$Y_{X/S}$	0.5 mg biomass formed/mg substrate consumed	

The volumetric recycle flow rate to volumetric feed flow rate ratio is

$$R \equiv \frac{F_7}{F_1} = 3.2$$

25. The sludge age, θ_X (biomass solids retention time or biomass in the bioreactor divided by the total biomass effluent rate), is most nearly

(A) 1 d
(B) 3 d
(C) 5 d
(D) 10 d

26. The bioreactor volume, V, is most nearly

(A) 100 L
(B) 600 L
(C) 800 L
(D) 1100 L

27. The biomass concentration, X_3, in the bioreactor is most nearly

(A) 540 mg biomass/L
(B) 1800 mg biomass/L
(C) 3200 mg biomass/L
(D) 5400 mg biomass/L

Problems 28–30 are based on the following information.

A continuous reactor is to be controlled at a steady-state operating point as illustrated.

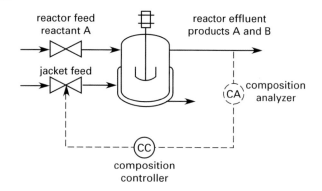

The reaction is first order, irreversible, exothermic, and takes place in a well-mixed stirred tank reactor. The reaction is

$$A \rightarrow B$$

The reactor is jacketed and constant volume, and the feed is pure A. The material and energy balances for this reactor are

$$\frac{dC_A}{dt} = \left(\frac{F}{V}\right)(C_{Af} - C_A) - k_o e^{\frac{-E}{RT}} C_A$$

$$\frac{dT}{dt} = \left(\frac{F}{V}\right)(T_f - T) + \left(\frac{(-\Delta H)}{\rho c_p}\right) k_o e^{\frac{-E}{RT}} C_A$$
$$\quad - \left(\frac{UA}{V \rho c_p}\right)(T - T_j)$$

The reactor conditions are

V	reactor volume, 1 m^3
F	volumetric feed rate, 0.1 m^3/s
C_{Af}	concentration of A in feed, 10 kmol/m^3
k_o	preexponential factor, $7.295 \times 10^5 s^{-1}$
E	activation energy, 8.3736×10^7 J/kmol
R	universal gas constant, 8314.510 J/kmol·K
$(-\Delta H)$	heat of reaction, 1.6747×10^8 J/kmol
ρ	reactor density, 1000 kg/m^3
c_p	reactor heat capacity, 5000 J/kg·K
T_f	500K
T_j	479.38K
A	1 m^2
U	1.0425×10^5 J/s·m^2·K

With these parameters, the reactor has three steady-state operating points for reactor temperature and concentration of A in the reactor.

operating point	Ts (K)	C_{As} (kmol/m^3)
1	500.01	9.8712
2	640.51	4.8020
3	750.01	0.8511

It is desired to operate at point 2, which has a higher productivity than point 1. (Point 3 has a higher productivity than point 2, but the product decomposes at that temperature.) This will be accomplished by manipulating jacket temperature to control the concentration of A.

28. The transfer function, $(C_A(s)[\text{kmol/m}^3])/(T_j(s)[\text{K}])$, between the jacket temperature input and the output concentration of A is most nearly

(A) $\dfrac{2.7 \times 10^{-4}}{s^2 + 0.098s - 0.018}$

(B) $\dfrac{-2.7 \times 10^{-4}}{s^2 - 0.098s - 0.018}$

(C) $\dfrac{-2.7 \times 10^{-4}}{s^2 + 0.098s - 0.018}$

(D) $\dfrac{-2.7 \times 10^{-4}}{s^2 + 0.098s + 0.018}$

29. What statement best describes the open-loop stability of the reactor at operating point 2?

(A) The system is stable.
(B) The system is unstable.
(C) The system is stable if the jacket temperature is steady.
(D) The system is stable if the feed concentration is steady.

30. It is proposed to use a proportional plus derivative controller to operate at point 2. The controller will be used to manipulate jacket temperature to control concentration. What statement best describes the closed-loop stability of the reactor at operating point 2?

(A) The system is stable for all $K < 0$m^3·K/kmol, $KT_D < 0$m^3·K·s/kmol.
(B) The system is unstable for all $K < 0$m^3·K/kmol, $KT_D < 0$m^3·K·s/kmol.
(C) The system is stable for all $K < -66$m^3·K/kmol, $KT_D > -370$m^3·K·s/kmol.
(D) The system is stable for all $K < -66$m^3·K/kmol, $KT_D < -370$m^3·K·s/kmol.

Problems 31–33 are based on the following information.

It is desired to estimate the optimal economic pipe insulation thickness for a 0.08 m outer diameter steel pipe carrying steam at 650K. The ambient indoor temperature is 300K and the air is still.

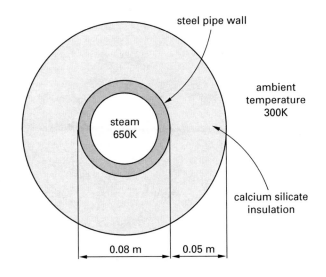

steel pipe wall

ambient temperature 300K

steam 650K

calcium silicate insulation

0.08 m   0.05 m

For aluminum jacketed calcium silicate insulation at a mean temperature of 500K, the thermal conductivity and emissivity are

k 0.073 W/m·K
ε 0.05

For air at a mean temperature of 325K, the density, heat capacity, viscosity, and thermal conductivity are

ρ 1.088 kg/m^3
c_p 1007 J/kg·K
μ 1.95 × 10^{-5} kg/s·m
k 0.0282 W/m·K

The pipe inside heat-transfer coefficient and pipe wall thermal conductivity may be assumed to be very large. Energy costs are \$10/GJ, and it is assumed the pipe is operated 8760 h/yr. The initial installed cost of insulation per meter of pipe length is a function of inside insulation diameter, D_i, and insulation thickness, t, given by the correlation

$$\text{installed cost}\left[\frac{\$}{\text{m}}\right] = 2340.4\left(D_i\,[\text{m}]\right)^{0.6}\left(t\,[\text{m}]\right)$$

The pretax minimum attractive rate of return is 50% per year and insulation is expected to last 10 yr before replacement.

31. If the insulation thickness is 0.05 m as illustrated, the outside overall heat-transfer coefficient is most nearly

(A) 5.6 W/m^2·K
(B) 12 W/m^2·K
(C) 23 W/m^2·K
(D) 42 W/m^2·K

32. The annual cost of the heat lost per linear meter of pipe is most nearly

(A) \$11/yr
(B) \$15/yr
(C) \$33/yr
(D) \$53/yr

33. If the outside heat-transfer coefficient is assumed constant for these conditions, the optimal insulation thickness is most nearly

(A) 0.05 m
(B) 0.08 m
(C) 0.11 m
(D) 0.14 m

Problems 34–36 are based on the following information.

Product B is to be produced from reactant A in a continuous stirred tank reactor (CSTR). The reaction is

$$A \rightarrow B$$

The feed concentration of A is 1000 mol/m^3 and the second-order rate coefficient is 0.00001 m^3/s·mol. The production rate is to be 1 million moles of B per year, assuming 8760 h of production per year. The unit cost of raw material A is \$0.10/mol and the investment cost of the reactor is \$100,000 /m^3. Assume an interest rate of 10% and a 10 yr equipment life.

34. The optimal conversion is most nearly

(A) 78%
(B) 84%
(C) 91%
(D) 97%

35. The optimal reactor volume is most nearly

(A) 0.1 m^3
(B) 0.4 m^3
(C) 3 m^3
(D) 7 m^3

36. The equivalent uniform annual cost (EUAC) at these conditions is most nearly

(A) \$10,000/yr
(B) \$50,000/yr
(C) \$90,000/yr
(D) \$100,000/yr

Problems 37–39 are based on the following information.

Water at 15°C is flowing from the bottom of a large diameter open tank through a piping system as illustrated.

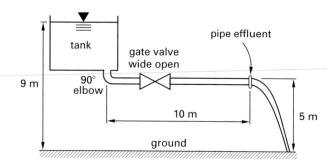

The water exits the tank bottom into a 90° elbow (0.9 velocity heads), through a wide open gate valve (0.2 velocity heads), through 10 m of nominal 2 in schedule 40 commercial steel pipe with inner diameter 2.067 in (0.0525 m), and then exits into the open. The water surface in the tank is 9 m above the ground and the exit is 5 m above the ground.

37. The average velocity in the pipe is most nearly

(A) 1.7 m/s
(B) 3.2 m/s
(C) 5.4 m/s
(D) 8.7 m/s

38. The volumetric flow rate in the pipe is most nearly

(A) 2.7 m^3/h
(B) 5.5 m^3/h
(C) 12 m^3/h
(D) 25 m^3/h

39. In a proposal to add water to the tank, a pump (efficiency 70%) is put in the line.

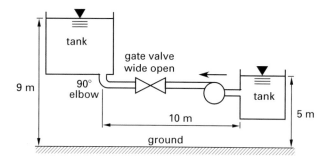

The suction of the pump draws from another large tank via a pipe with liquid surface at 5 m above the ground. The surface of the receiving tank is again 9 m above the ground. The pipe length is again 10 m of nominal 2 in schedule 40 commercial steel pipe with a 2.067 in (0.0525 m) inner diameter. The desired flow rate is 0.025 m^3/s. The required power of the pump is most nearly

(A) −16 kW
(B) −20 kW
(C) −24 kW
(D) +32 kW

Problems 40 and 41 are based on the following information.

In a flow process to make polyethylene at high temperatures and pressures, ethylene decomposes violently into essentially carbon and methane with a minor amount of hydrogen and hydrocarbons. The overall stoichiometry may be assumed to be

$$C_2H_4(g) \rightarrow C(s) + CH_4(g)$$

Assume the reactor is adiabatic and filled with pure ethylene at a temperature of 500°C and a pressure of 2000 bar (200 MPa) with complete conversion.

component	state	$\Delta \hat{H}_f^o$ (298.15K) (J/mol)	c_p^o (J/mol·K)
ethylene	gas	52 510	28.161 + 0.054 016T [K]
carbon	solid	0	5.2173 + 0.011 073T [K]
methane	gas	−74 520	22.192 + 0.043 151T [K]

40. The temperature after the decomposition is most nearly

(A) 800K
(B) 1000K
(C) 2000K
(D) 2500K

41. Neglecting the solid carbon volume, the pressure after the decomposition is most nearly

(A) 2000 bar (200 MPa)
(B) 3000 bar (300 MPa)
(C) 5000 bar (500 MPa)
(D) 6000 bar (600 MPa)

Problems 42 and 43 are based on the following information.

A 1 cm diameter cylinder of naphthalene is in an air stream with the flow perpendicular to the axis of the cylinder. The air stream has a velocity of 1 m/s and a temperature of 20°C. For flow perpendicular to the axis of a constant-concentration circular cylinder, the following mass-transfer correlation has been found from analogy with the corresponding heat-transfer situation.

$$\text{Sh} = 0.43 + 0.532 \, (\text{Re})^{0.5} \, (\text{Sc})^{0.31} \quad [1 < \text{Re} < 4000]$$

	Sherwood number	$\dfrac{k_G\,(p_B)_{\mathrm{lm}}\,DRT}{PD_m}$
Sh	Sherwood number	
k_G	mass-transfer coefficient	$\left[\dfrac{\text{mol}}{\text{m}^2\text{·s·Pa}}\right]$
$(p_B)_{\mathrm{lm}}$	$(p_{B2}0 - p_{B1})/(\ln(p_{B2}/p_{B1}))$, log mean partial pressure of component B	[Pa]
D	cylinder diameter	[m]
R	universal gas constant	8.314 Pa·m^3/ mol·K
T	temperature	[K]
P	pressure	[Pa]
D_m	diffusivity	$\left[\dfrac{\text{m}^2}{\text{s}}\right]$
Re	Reynolds number	$\dfrac{DV\rho}{\mu}$
Sc	Schmidt number	$\dfrac{\mu}{\rho D_m}$

The physical properties of air are

T (°C)	viscosity (Pa·s)	density (kg/m^3)
20	1.8×10^{-5}	1.2056

The physical properties of naphthalene are

T (°C)	diffusion coefficient (m^2/s)	vapor pressure (Pa)
20	6.2×10^{-6}	6.67

42. The mass-transfer coefficient is most nearly

(A) 5×10^{-6} mol/m^2·s·Pa
(B) 5×10^{-4} mol/m^2·s·Pa
(C) 5×10^{-3} mol/m^2·s·Pa
(D) 5×10^{-2} mol/m^2·s·Pa

43. The molar flux of napthalene away from the cylinder is most nearly

(A) 3×10^{-5} mol/m^2·s
(B) 3×10^{-3} mol/m^2·s
(C) 3×10^{-2} mol/m^2·s
(D) 3×10^{-1} mol/m^2·s

44. From the given table of solubility-product data, which compound would dissolve most easily?

compound	formula	K_{sp}
cadmium sulfide	CdS	8×10^{-28}
copper sulfide	CuS	6×10^{-37}
magnesium hydroxide	Mg(OH)$_2$	1.6×10^{-12}
silver iodide	AgI	8.3×10^{-17}

(A) CdS
(B) CuS
(C) Mg(OH)$_2$
(D) AgI

45. If the hydrogen ion concentration (pH) in a solution is known to be 5.0×10^{-6} M, what is most nearly the pH of the solution?

(A) 0.70
(B) 5.3
(C) 7.5
(D) 12

46. The equilibrium constant of hypochlorus acid (HClO) is 3.2×10^{-8}. The concentrations of ClO and HClO are 1.5×10^{-5} and 0.7×10^{-7}, respectively. What is the concentration of the [H$^+$] ion?

(A) 1.5×10^{-10}
(B) 6.9×10^{-7}
(C) 1.5×10^{-6}
(D) 3.2×10^{-5}

47. What is the structural formula of sec-butyl alcohol?

(A)   CH$_3$—CH—CH$_2$—OH
                    |
                   CH$_3$

(B)   CH$_3$—CH$_2$—CH—OH
                          |
                         CH$_3$

(C)   CH$_3$—CH—OH
                    |
                   CH$_3$

(D)   CH$_3$—OH

48. Which of the following equations is NOT balanced?

(A) $(CH_3)_3COH + HCl \rightarrow (CH_3)_3CCl + H_2O$
(B) $C_5H_{11}Na + CH_3CO_2H \rightarrow C_5H_{12} + CH_3CO_2Na$
(C) $2\ KClO_3 \rightarrow 2\ KCl + 3\ O_2$
(D) $2\ C_6H_{12} + 12\ O_2 \rightarrow 6\ CO_2 + 6\ CO + 12\ H_2O$

49. Select the appropriate name for the compound shown.

(A) toluene
(B) o-xylene
(C) m-xylene
(D) p-xylene

Problems 50–53 are based on the following information.

It is desired to remove 95% of the propylene from a gas stream composed of propane, propylene, butane, and pentane. A distillation tower is used to make this recovery. The compositions of the feed and bottom product streams are given in the following table.

component	feed (wt%)	bottoms (wt%)
propane	20	10
propylene	25	5
butane	35	15
pentane	20	70

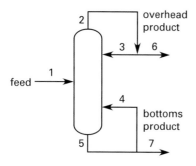

50. If the feed stream mass flow rate is 1500 kg/h, what is most nearly the total mass flow rate of the bottoms stream?

(A) 300 kg/h
(B) 370 kg/h
(C) 410 kg/h
(D) 450 kg/h

51. What is most nearly the percent propylene in the overhead product stream given the flow rates of the feed stream and bottoms streams are 1500 kg/h and 1000 kg/h, respectively?

(A) 50%
(B) 65%
(C) 75%
(D) 80%

52. The reflux ratio is 2:1, and the flow rate of the column bottoms stream 5 is four times the flow rate of the bottoms product stream 7. The specific enthalpies of the various streams are given in the following table.

stream	specific enthalpy (kcal/kg)
1	?
2	90
3	60
4	182.5
5	100

What is most nearly the specific enthalpy of the feed stream 1, given the flow rates of stream 1 and stream 7 are 3000 kg/h and 800 kg/h, respectively?

(A) 71 kcal/kg
(B) 100 kcal/kg
(C) 110 kcal/kg
(D) 130 kcal/kg

53. If the bottoms product stream 7 has a mass flow rate of 500 kg/h with no change in the bottoms composition, what will be the resulting composition of propylene in the overhead product stream 6? Use a feed flow rate of 1500 kg/h.

(A) 20%
(B) 25%
(C) 30%
(D) 35%

54. A pump currently delivers 250 m³/h of a fluid using an impeller with a 200 mm diameter. If the impeller diameter is increased to 230 mm and the pump speed is held constant, what is most likely the new volumetric flow rate of the fluid?

(A) 270 m³/h
(B) 290 m³/h
(C) 330 m³/h
(D) 380 m³/h

Problems 55 and 56 are based on the following information.

A fluid supplies a compound pipe system at 1000 m³/h, as shown. The length of branch 3 is 50% more than the length of branch 2. Branch 3 has an internal diameter of 150 mm, while branch 2 is has an internal diameter of 100 mm.

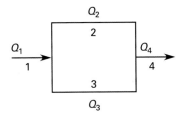

55. If the friction factors in branches 2 and 3 are about the same. What is most nearly the volumetric flow rate in branch 2?

(A) 270 m³/h
(B) 310 m³/h
(C) 450 m³/h
(D) 500 m³/h

56. If the friction factors are 0.015 in branch 2 and 0.025 in branch 3, what is most nearly the volumetric flow rate in branch 2?

(A) 290 m³/h
(B) 320 m³/h
(C) 370 m³/h
(D) 410 m³/h

57. A typical process flow diagram usually contains several items describing the process, such as a simplified sketch of the system, a mass balance, pressures, and temperatures. Which of the following is also normally found on a process flow diagram?

(A) line sizes
(B) emergency shower locations
(C) major control elements
(D) exchanger tube sizes

58. A new process unit is being studied for possible installation. A 25 ton/yr unit would cost $55 million (M) to install. Based on the following critical equipment costs and scale factors, what would most likely be the cost of a new 50 ton/yr unit?

equipment	cost ($)	scale factor exponent
exchangers	10 M	0.55
vessels	15 M	0.50
pumps	10 M	0.60
piping	15 M	0.70
miscellaneous	5 M	0.75
total	55 M	

(A) $76 M
(B) $84 M
(C) $86 M
(D) $93 M

59. Of the following operations, which is performed first?

(A) multiplication
(B) division
(C) exponentiation
(D) addition

60. A group of five product silos was tested for additive levels. The following results were obtained.

silo 1: 2300 ppm
silo 2: 2800 ppm
silo 3: 2500 ppm
silo 4: 2600 ppm
silo 5: 3100 ppm

The following equation defines standard deviation.

$$\sigma = \sqrt{\frac{\sum(X - \overline{X})^2}{n - 1}}$$

What is most nearly the standard deviation for the given additive level results?

(A) 160 ppm
(B) 310 ppm
(C) 330 ppm
(D) 540 ppm

SOLUTIONS

1. The first-order rate constant is given by

$$\ln \frac{C_A}{C_{Ao}} = -kt$$

The half-life is the time it takes for the reactant to decay to half of its initial concentration. The rate constant can be calculated from the half-life,

$$k = \frac{-\ln \frac{C_A}{C_{Ao}}}{t} = \frac{-\ln \frac{1}{2}}{t_{1/2}} = \frac{\ln 2}{t_{1/2}}$$

$$= \frac{0.6931}{1000 \text{ s}}$$

$$= 0.000\,693\,1 \text{ s}^{-1} \quad (0.0007 \text{ s}^{-1})$$

The answer is C.

2. Calculate the design equation for a batch reactor with first-order kinetics.

$$\ln(1 - X_A) = -kt$$

Solving for the batch time,

$$t = \frac{-\ln(1 - X_A)}{k} = \frac{-\ln(1 - 0.90)}{0.000\,693\,1 \text{ s}^{-1}}$$

$$= 3322 \text{ s}$$

The reactor volume is related to the batch time and production rate by

$$V_{\text{batch}} = \frac{t F_{Ao}}{C_{Ao}}$$

$$= (3322 \text{ s}) \left(1000 \, \frac{\text{kg}}{\text{h}} \right) \left(\frac{1 \text{ h}}{3600 \text{ s}} \right) \left(1000 \, \frac{\text{g}}{\text{kg}} \right)$$

$$\times \left(\frac{1 \text{ mol}}{100 \text{ g}} \right) \left(\frac{1 \text{ L}}{1 \text{ mol}} \right) \left(\frac{1 \text{ m}^3}{1000 \text{ L}} \right)$$

$$= 9.2 \text{ m}^3 \quad (10 \text{ m}^3)$$

The answer is B.

3. Calculate the design equation for a CSTR.

$$\frac{\tau}{C_{Ao}} = \frac{V_{\text{CSTR}}}{F_{Ao}} = \frac{X_A}{-r_A}$$

Solving for the space-time for a first-order reaction,

$$\tau = \frac{C_{Ao} X_A}{-r_A} = \frac{C_{Ao} X_A}{k C_A}$$

$$= \frac{C_{Ao} X_A}{k C_{Ao}(1 - X_A)}$$

$$= \frac{X_A}{k(1 - X_A)}$$

$$= \frac{0.90}{(0.000\,693\,1 \text{ s}^{-1})(1 - 0.90)}$$

$$= 12\,985 \text{ s}$$

Relate the reactor volume to the space-time and feed rate.

$$V_{\text{CSTR}} = \frac{\tau F_{Ao}}{C_{Ao}}$$

$$= (12\,985 \text{ s}) \left(1000 \, \frac{\text{kg}}{\text{h}} \right) \left(\frac{1 \text{ h}}{3600 \text{ s}} \right)$$

$$\times \left(1000 \, \frac{\text{g}}{\text{kg}} \right) \left(\frac{1 \text{ mol}}{100 \text{ g}} \right) \left(\frac{1 \text{ L}}{1 \text{ mol}} \right) \left(\frac{1 \text{ m}^3}{1000 \text{ L}} \right)$$

$$= 36.069 \text{ m}^3 \quad (40 \text{ m}^3)$$

The answer is C.

4. The total batch time is

$$t = \frac{-\ln(1 - X_A)}{k} + t_{\text{downtime}}$$

$$= \frac{-\ln(1 - 0.90)}{(0.000\,693\,1 \text{ s}^{-1})} + (3 \text{ h}) \left(3600 \, \frac{\text{s}}{\text{h}} \right)$$

$$= 14\,122 \text{ s}$$

The reactor volume is related to the batch time and production rate by

$$V_{\text{batch}} = \frac{t F_{Ao}}{C_{Ao}}$$

$$= (14\,122 \text{ s}) \left(\frac{1000 \text{ kg}}{1 \text{ h}} \right) \left(\frac{1 \text{ h}}{3600 \text{ s}} \right) \left(\frac{1000 \text{ g}}{1 \text{ kg}} \right)$$

$$\times \left(\frac{1 \text{ mol}}{100 \text{ g}} \right) \left(\frac{1 \text{ L}}{1 \text{ mol}} \right) \left(\frac{1 \text{ m}^3}{1000 \text{ L}} \right)$$

$$= 39.22 \text{ m}^3$$

$$V_{\text{batch}} > V_{\text{CSTR}}$$

The answer is B.

5. Solving for the space-time for a first-order reaction.

$$\tau_{N-\text{reactors}} = \left(\frac{N}{k} \right) \left(\left(\frac{C_{Ao}}{C_{AN}} \right)^{1/N} - 1 \right)$$

$$= \left(\frac{N}{k} \right) \left(\left(\frac{C_{Ao}}{C_{Ao}(1 - X_A)} \right)^{1/N} - 1 \right)$$

$$= \left(\frac{N}{k} \right) \left(\left(\frac{1}{1 - X_A} \right)^{1/N} - 1 \right)$$

$$\tau_{5-\text{reactors}} = \left(\frac{5}{0.000\,693\,1 \text{ s}^{-1}} \right) \left(\left(\frac{1}{1 - 0.90} \right)^{1/5} - 1 \right)$$

$$= 4219.4 \text{ s}$$

Relate the reactor volume to the space-time and feed rate.

$$V_{5\text{-reactors}} = \frac{\tau_{5\text{-reactors}} F_{Ao}}{C_{Ao}}$$

$$= (4219.4 \text{ s}) \left(\frac{1000 \text{ kg}}{1 \text{ h}} \right) \left(\frac{1 \text{ h}}{3600 \text{ s}} \right) \left(\frac{1000 \text{ g}}{1 \text{ kg}} \right)$$

$$\times \left(\frac{1 \text{ mol}}{100 \text{ g}} \right) \left(\frac{1 \text{ L}}{1 \text{ mol}} \right) \left(\frac{1 \text{ m}^3}{1000 \text{ L}} \right)$$

$$= 11.7 \text{ m}^3 \quad (12 \text{ m}^3)$$

The answer is C.

6. For the plug-flow reactor (PFR) equation and first order kinetics,

$$\tau = \frac{C_{Ao} V_{\text{PFR}}}{F_{Ao}} = C_{Ao} \int_0^{X_A} \frac{dX_A}{-r_A}$$

$$= C_{Ao} \int_0^{X_A} \frac{dX_A}{kC_A}$$

The volume of the reacting mass is assumed to vary with conversion according to

$$V = V_{X_A=0} \left(1 + \varepsilon_A X_A \right)$$

$$\varepsilon_A = \frac{V_{X_A=1} - V_{X_A=0}}{V_{X_A=0}}$$

The concentration must be related to the conversion. Concentration for a variable volume system is

$$C_A = \frac{N_A}{V} = \frac{N_{Ao}(1 - X_A)}{V_{X_A=0}(1 + \varepsilon_A X_A)}$$

$$= C_{Ao} \left(\frac{1 - X_A}{1 + \varepsilon_A X_A} \right)$$

Insert this into the PFR equation to obtain the result,

$$\tau = C_{Ao} \int_0^{X_A} \frac{dX_A}{kC_A}$$

$$= C_{Ao} \int_0^{X_A} \frac{(1 + \varepsilon_A X_A) dX_A}{k C_{Ao}(1 - X_A)}$$

$$k\tau = \int_0^{X_A} \frac{(1 + \varepsilon_A X_A) dX_A}{1 - X_A}$$

The answer is D.

Notice that (B) is the form for a constant-volume, irreversible PFR.

7. The heat of formation of the reaction at 25°C is given by

$$\Delta H_r^\circ = \sum_{\text{products}} \nu_i \left(\Delta \hat{H}_f^\circ \right)_i - \sum_{\text{reactants}} \nu_i \left(\Delta \hat{H}_f^\circ \right)_i$$

$$= -94\,390 \, \frac{\text{cal}}{\text{mol}} - \left(-70\,940 \, \frac{\text{cal}}{\text{mol}} + \left(\frac{1}{2} \right) (0) \right)$$

$$= -23\,450 \text{ cal/mol}$$

The free energy of formation of the reaction at 25°C is

$$\Delta G_r^\circ = \sum_{\text{products}} \nu_i \left(\Delta \hat{G}_f^\circ \right)_i - \sum_{\text{reactants}} \nu_i \left(\Delta \hat{G}_f^\circ \right)_i$$

$$= -88\,590 \, \frac{\text{cal}}{\text{mol}} - \left(-71\,680 \, \frac{\text{cal}}{\text{mol}} + \left(\frac{1}{2} \right) (0) \right)$$

$$= -16\,910 \text{ cal/mol}$$

The equilibrium constant at 25°C is given by

$$\Delta G = -RT \ln K_a$$

$$K_a = e^{-\frac{\Delta G}{RT}}$$

$$= e^{-\frac{-16\,910 \, \frac{\text{cal}}{\text{mol}}}{\left(1.9872 \, \frac{\text{cal}}{\text{mol·K}} \right)(25°\text{C}+273°)}}$$

$$= 2.4839 \times 10^{12}$$

Assuming the enthalpy of formation is constant at any temperature,

$$\frac{d \ln K}{dT} = \frac{\Delta H^\circ}{RT^2}$$

$$\int_{K_1}^{K_2} d \ln K = \int_{T_1}^{T_2} \frac{\Delta H^\circ}{RT^2} dT = \frac{\Delta H^\circ}{R} \int_{T_1}^{T_2} \frac{dT}{T^2}$$

$$\ln \frac{K_2}{K_1} = \left(-\frac{\Delta H^\circ}{R} \right) \left(\frac{1}{T_2} - \frac{1}{T_1} \right)$$

$$K_2 = K_1 \left[e^{-\frac{\Delta H^\circ}{R} \left(\frac{1}{T_2} - \frac{1}{T_1} \right)} \right]$$

$$= 2.4839 \times 10^{12}$$

$$\times \left[e^{-\left(\frac{-23\,450 \, \frac{\text{cal}}{\text{mol}}}{1.9872 \, \frac{\text{cal}}{\text{mol·K}}} \right) \left(\frac{1}{538°\text{C}+273°} - \frac{1}{25°\text{C}+273°} \right)} \right]$$

$$= 33.435$$

The free energy at 538°C is given by

$$\Delta G = -RT \ln K$$

$$= - \left(1.9872 \, \frac{\text{cal}}{\text{mol·K}} \right) (538\text{K} + 273.15\text{K})$$

$$\times \ln 33.435$$

$$= -5657.2 \text{ cal/mol} \quad (-5700 \text{ cal/mol})$$

The answer is A.

8. The mole table for this problem is

species	moles initial	moles react	moles final	mole fraction
SO_2	1	$-x$	$1-x$	$\dfrac{1-x}{\frac{3}{2}-\frac{x}{2}}$
O_2	$\dfrac{1}{2}$	$-\dfrac{x}{2}$	$\dfrac{1}{2}-\dfrac{x}{2}$	$\dfrac{\frac{1}{2}-\frac{x}{2}}{\frac{3}{2}-\frac{x}{2}}$
SO_3	0	x	x	$\dfrac{x}{\frac{3}{2}-\frac{x}{2}}$
total	$\dfrac{3}{2}$	$-\dfrac{x}{2}$	$\dfrac{3}{2}-\dfrac{x}{2}$	1

The equilibrium constant is given from the following equation and $f_i^o = 1$ bar.

$$K_p = \frac{\dfrac{p_{SO_3}}{f_{SO_3}^o}}{\left(\dfrac{p_{SO_2}}{f_{SO_2}^o}\right)\left(\dfrac{p_{O_2}}{f_{O_2}^o}\right)^{1/2}} = \frac{y_{SO_3}}{y_{SO_2}y_{O_2}^{1/2}\left(\dfrac{P}{f_i^o}\right)^{1/2}}$$

$$= \frac{\dfrac{x}{\frac{3}{2}-\frac{x}{2}}}{\left(\dfrac{1-x}{\frac{3}{2}-\frac{x}{2}}\right)\left(\dfrac{\frac{1}{2}-\frac{x}{2}}{\frac{3}{2}-\frac{x}{2}}\right)^{1/2}\left(\dfrac{5\text{ bar}}{1\text{ bar}}\right)^{1/2}}$$

$$33.435 = \frac{x}{(1-x)\left(\dfrac{\frac{1}{2}-\frac{x}{2}}{\frac{3}{2}-\frac{x}{2}}\right)^{1/2}(5)^{1/2}}$$

The solution to this equation by the method of trial and error is the conversion

$$x = 0.931\,52$$
$$= (0.931\,52) \times 100\%$$
$$= 93.152\% \quad (93\%)$$

The answer is D.

9. The gas constant in the units of the problem is

$$R = \left(8.314\ \frac{\text{kPa·m}^3}{\text{kmol·K}}\right)\left(\frac{1\text{ kmol}}{1000\text{ mol}}\right)$$
$$\times \left(\frac{1\text{ bar}}{100\text{ kPa}}\right)\left(100\ \frac{\text{cm}}{\text{m}}\right)^3$$
$$= 83.14\ \frac{\text{bar·cm}^3}{\text{mol·K}}$$

The constants in the equation are

$$a = \frac{0.427\,48R^2T_c^{2.5}}{p_c}$$

$$= \frac{(0.427\,48)\left(83.14\ \dfrac{\text{bar·cm}^3}{\text{mol·K}}\right)^2(510.0\text{K})^{2.5}}{60.8\text{ bars}}$$

$$= 2.8547 \times 10^8\ \frac{\text{cm}^6\text{ bar K}^{1/2}}{\text{mol}^2}$$

$$b = \frac{0.086\,64RT_c}{p_c}$$

$$= \frac{0.086\,64\left(83.14\ \dfrac{\text{bar·cm}^3}{\text{mol·K}}\right)(510.0\text{K})}{60.8\text{ bars}}$$

$$= 60.422\text{ cm}^3/\text{mol}$$

The temperature and pressure conditions are

$$T = 60°C + 273.15°$$
$$= 333.15\text{K}$$
$$P = 10\text{ bars}$$

The Redlich-Kwong equation of state is

$$V = \frac{RT}{P} + b - \frac{a(V-b)}{T^{1/2}PV(V+b)}$$

$$= \frac{\left(83.14\ \dfrac{\text{bar·cm}^3}{\text{mol·K}}\right)(333.15\text{K})}{10\text{ bars}}$$
$$+ 60.422\ \frac{\text{cm}^3}{\text{mol}}$$

$$- \frac{\left(2.8547 \times 10^8\ \dfrac{\text{cm}^6\text{·bar·K}^{1/2}}{\text{mol}^2}\right)\times\left(V - 60.422\ \dfrac{\text{cm}^3}{\text{mol}}\right)}{(333.15\text{K})^{1/2}(10\text{ bar})V\left(V + 60.422\ \dfrac{\text{cm}^3}{\text{mol}}\right)}$$

Or in terms of pure numbers, the molar volume is

$$V = 2830.2 - \frac{(1.564 \times 10^6)(V - 60.422)}{V(V+60.422)}\ \frac{\text{cm}^3}{\text{mol}}$$

This is in a form useful for the method of successive substitution, which takes the form of the following iterative equation.

$$V_{i+1} = fV_i$$

Use an initial guess from the ideal gas law and iterate. The gas molar volume guess is

$$V_o = \frac{\left(83.14\ \dfrac{\text{bar·cm}^3}{\text{mol·K}}\right)(333.15\text{K})}{10\text{ bars}}$$
$$= 2769.8\text{ cm}^3/\text{mol}$$

The iteration history is

i	V_i (cm^3/mol)
0	2769.8
1	2289.6
2	2282.2
3	2152.1
4	2143.2
5	2140.5
6	2139.6
7	2139.4
8	2139.3

The molar volume is

$$V = 2139.3 \text{ cm}^3/\text{mol} \quad (2100 \text{ cm}^3/\text{mol})$$

The answer is C.

10. The constant, c, becomes

$$c = b^2 + \frac{bRT}{P} - \frac{a}{PT^{1/2}}$$

$$= \left(60.422 \ \frac{\text{cm}^3}{\text{mol}}\right)^2$$

$$+ \frac{\left(60.422 \ \frac{\text{cm}^3}{\text{mol}}\right)\left(83.14 \ \frac{\text{bar·cm}^3}{\text{mol·K}}\right)(333.15\text{K})}{10 \text{ bars}}$$

$$- \frac{\left(2.8547 \times 10^8 \ \frac{\text{cm}^6\text{·bar·K}^{1/2}}{\text{mol}^2}\right)}{(10 \text{ bars})(333.15\text{K})^{1/2}}$$

$$= -1.393 \times 10^6 \text{ cm}^6/\text{mol}^2$$

The *R-K* equation becomes

$$V = \frac{V^3 - \frac{RT}{P}V^2 - \frac{ab}{PT^{1/2}}}{c}$$

$$= \frac{\left(V^3 - \frac{\left(83.14 \ \frac{\text{bar·cm}^3}{\text{mol·K}}\right)(333.15\text{K})}{10 \text{ bars}}V^2 \atop - \frac{\left(60.422 \ \frac{\text{cm}^3}{\text{mol}}\right)\left(2.8547 \times 10^8 \ \frac{\text{cm}^6\text{·bar·K}^{1/2}}{\text{mol}^2}\right)}{(10 \text{ bars})(333.15\text{K})^{1/2}}\right)}{-1.393 \times 10^6 \ \frac{\text{cm}^6}{\text{mol}^2}}$$

$$= \frac{V^3 - 2769.8V^2 - 9.4501 \times 10^7}{-1.393 \times 10^6}$$

The iteration history starting with b as a first guess is

i	V_i (cm^3/mol)
0	60.422
1	74.941
2	78.705
3	79.807
4	80.139
5	80.240
6	80.271

The liquid molar volume is

$$V = 80.271 \text{ cm}^3/\text{mol} \quad (80 \text{ cm}^3/\text{mol})$$

The answer is C.

Since this is a cubic equation of state, there are actually three roots, the third being $V = 550 \text{ cm}^3/\text{mol}$, which does not correspond to any real physical density of the system.

11. The detailed calculations given by the pseudocode are in order.

$$X = 3$$
$$J = 1$$
$$Y = \frac{2 - 2\sin x}{3} = \frac{2 - 2\sin 3}{3}$$
$$= 0.572\,59$$
$$\text{XC} = \frac{3 - 2\cos y}{3} = \frac{3 - 2\cos 0.572\,59}{3}$$
$$= 0.439\,67$$
$$\text{ERROR} = \frac{\text{XC} - X}{X} = \frac{0.439\,67 - 3}{3}$$
$$= -0.853\,44$$

$$X = \text{XC} = 0.439\,67$$
$$J = 2$$
$$Y = \frac{2 - 2\sin x}{3} = \frac{2 - 2\sin 0.439\,67}{3}$$
$$= 0.382\,91$$
$$\text{XC} = \frac{3 - 2\cos y}{3} = \frac{3 - 2\cos 0.382\,91}{3}$$
$$= 0.381\,61$$
$$\text{ERROR} = \frac{\text{XC} - X}{X} = \frac{0.381\,61 - 0.439\,67}{0.439\,67}$$
$$= -0.132\,05$$

$$X = \text{XC} = 0.381\,61$$
$$J = 3$$

$$Y = \frac{2 - 2\sin x}{3} = \frac{2 - 2\sin 0.381\,61}{3}$$
$$= 0.418\,39$$
$$XC = \frac{3 - 2\cos y}{3} = \frac{3 - 2\cos 0.418\,39}{3}$$
$$= 0.390\,84$$
$$\text{ERROR} = \frac{XC - X}{X} = \frac{0.390\,84 - 0.381\,61}{0.381\,61}$$
$$= 0.024\,187$$

Stop here because ABS(0.024 187) < 0.1. In summary the iteration history is

J	X	Y	ERROR
1	3.000	0.573	−0.853
2	0.440	0.383	−0.132
3	0.382	0.418	0.0241

The answer is C.

12. The heat balance for the exchanger given in the Heat Transfer section in the NCEES Handbook is that the cold-side transfer duty equals the hot-side transfer duty equals the design equation for the exchanger.

$$Q = \dot{m}_C c_{pC}(T_{Co} - T_{Ci})$$
$$Q = \dot{m}_H c_{pH}(T_{Hi} - T_{Ho})$$
$$Q = UA\Delta T_{lm}$$

The cold-side duty from the cold-side energy balance is

$$Q = \dot{m}_C c_{pC}(T_{Co} - T_{Ci})$$
$$= \left(1.26\,\frac{\text{kg}}{\text{s}}\right)\left(2500\,\frac{\text{J}}{\text{kg·K}}\right)$$
$$\times (40°C - 20°C)\left(1\,\frac{\text{K}}{°C}\right)$$
$$= 63\,000 \text{ J/s}$$

The hot water flow can then be obtained from the hot-side energy balance.

$$\dot{m}_H = \frac{Q}{c_{pH}(T_{Hi} - T_{Ho})}$$
$$= \frac{63\,000\,\frac{\text{J}}{\text{s}}}{\left(4200\,\frac{\text{J}}{\text{kg·K}}\right)(80°C - 50°C)\left(1\,\frac{\text{K}}{°C}\right)}$$
$$= 0.50 \text{ kg/s}$$

The answer is A.

13. The log mean temperature difference (LMTD) for countercurrent flow in the Heat Transfer section of the NCEES Handbook is

$$\Delta T_{lm} = \frac{(T_{Ho} - T_{Ci}) - (T_{Hi} - T_{Co})}{\ln\left(\frac{T_{Ho} - T_{Ci}}{T_{Hi} - T_{Co}}\right)}$$
$$= \frac{(50°C - 20°C) - (80°C - 40°C)}{\ln\left(\frac{50°C - 20°C}{80°C - 40°C}\right)}$$
$$= 34.76°C \quad (35°C)$$

The log mean temperature difference (LMTD) for concurrent flow in the Heat Transfer section of the NCEES Handbook is

$$\Delta T_{lm} = \frac{(T_{Ho} - T_{Co}) - (T_{Hi} - T_{Ci})}{\ln\left(\frac{T_{Ho} - T_{Co}}{T_{Hi} - T_{Ci}}\right)}$$
$$= \frac{(50°C - 40°C) - (80°C - 20°C)}{\ln\left(\frac{50°C - 40°C}{80°C - 20°C}\right)}$$
$$= 27.91°C \quad (28°C)$$

Choose the larger LMDT for smaller heat exchanger area and a lower cost exchanger.

The answer is B.

14. The cross-sectional area of the inside tube is

$$A_x = \frac{\pi D_i^2}{4}$$
$$= \left(\frac{\pi (5.250 \text{ cm})^2}{4}\right)\left(\frac{1 \text{ m}}{100 \text{ cm}}\right)^2$$
$$= 0.002\,164\,8 \text{ m}$$

The mass flow, Reynolds, and Prandtl numbers from the Heat Transfer section of the NCEES Handbook are
$$\dot{m} = \rho V A_x$$
$$\text{Re} = \frac{VD\rho}{\mu} = \frac{\dot{m}D}{A_x \mu}$$
$$= \frac{\left(1.26\,\frac{\text{kg}}{\text{s}}\right)(5.250 \text{ cm})\left(\frac{1 \text{ m}}{100 \text{ cm}}\right)}{(0.002\,164\,8 \text{ m}^2)(0.25 \text{ cp})\left(\frac{1 \text{ Pa·s}}{1000 \text{ cp}}\right)}$$
$$= 122\,230$$
$$\text{Pr} = \frac{c_p \mu}{k}$$
$$= \frac{\left(2500\,\frac{\text{J}}{\text{kg·K}}\right)(0.25 \text{ cp})\left(\frac{1 \text{ Pa·s}}{1000 \text{ cp}}\right)}{0.182\,\frac{\text{W}}{\text{m·K}}}$$
$$= 3.4341$$

For turbulent flow in a pipe, from the Heat Transfer section of the NCEES Handbook, the Nusselt number is

$$\mathrm{Nu} = \frac{hD}{k}$$

$$= 0.023(\mathrm{Re})^{0.8}(\mathrm{Pr})^{1/3}\left(\frac{\mu_b}{\mu_w}\right)^{0.14}$$

$$[\mathrm{Re} > 10^4, \ \mathrm{Pr} > 0.7]$$

Assuming there is no variation of viscosity with temperature, the wall and bulk viscosities may be assumed equal and the Nusselt number becomes

$$\mathrm{Nu} = (0.023)(122\,230)^{0.8}(3.4341)^{1/3}(1)^{0.14}$$

$$= 407.45$$

The inside heat-transfer coefficient is

$$h_i = \mathrm{Nu}\frac{\mathrm{k}}{\mathrm{D_i}}$$

$$= (407.45)\left(\frac{0.182\ \dfrac{\mathrm{W}}{\mathrm{m \cdot K}}}{(5.250\ \mathrm{cm})\left(\dfrac{1\ \mathrm{m}}{100\ \mathrm{cm}}\right)}\right)$$

$$= 1412.5\ \mathrm{W/m^2 \cdot K} \quad (1400\ \mathrm{W/m^2 \cdot K})$$

The answer is D.

15. The overall heat-transfer coefficient from the Heat Transfer section of the NCEES Handbook is

$$\frac{1}{UA} = \frac{1}{h_i A_i} + \frac{R_{fi}}{A_i} + \frac{t}{kA_{\mathrm{avg}}} + \frac{R_{fo}}{A_o} + \frac{1}{h_o A_o}$$

Write this in terms of inner tube outside area,

$$\frac{1}{U_o} = \frac{A_o}{h_i A_i} + \frac{R_{fi}A_o}{A_i} + \frac{tA_o}{kA_{\mathrm{avg}}} + R_{fo} + \frac{1}{h_o}$$

For thin-walled tubes this becomes, in terms of diameter,

$$\frac{1}{U_o} = \frac{D_o}{h_i D_i} + \frac{R_{fi}D_o}{D_i} + \frac{2tD_o}{k(D_i + D_o)} + R_{fo} + \frac{1}{h_o}$$

The outside overall heat-transfer coefficient, neglecting inside fouling and wall heat transfer resistance, becomes

$$U_o = \left(\frac{D_o}{h_i D_i} + \frac{R_{fi}D_o}{D_i} + \frac{2tD_o}{k(D_i + D_o)} + R_{fo} + \frac{1}{h_o}\right)^{-1}$$

$$= \left(\frac{6.0325\ \mathrm{cm}}{\left(1412.5\ \dfrac{\mathrm{W}}{\mathrm{m^2 \cdot K}}\right)(5.250\ \mathrm{cm})} + 0 + 0 + 0.000\,35\ \dfrac{\mathrm{m^2 \cdot K}}{\mathrm{W}} + \dfrac{1}{3200\ \dfrac{\mathrm{W}}{\mathrm{m^2 \cdot K}}}\right)^{-1}$$

$$= 677.51\ \mathrm{W/m^2 \cdot K} \quad (680\ \mathrm{W/m^2 \cdot K})$$

The outside heat-exchanger area for countercurrent flow from the Heat Transfer section of the NCEES Handbook is

$$A_o = \frac{Q}{U_o \Delta T_{\mathrm{lm}}}$$

$$= \frac{63\,000\ \dfrac{\mathrm{J}}{\mathrm{s}}}{\left(677.51\ \dfrac{\mathrm{W}}{\mathrm{m^2 \cdot K}}\right)(34.76\ ^\circ\mathrm{C})}$$

$$= 2.6751\ \mathrm{m} \quad (2.7\ \mathrm{m})$$

The answer is C.

16. The feed is 50 mol% acetone, so the mole fraction of acetone in the feed is 0.50. The acetone flow in the feed is

$$Fz_F = \left(1000\ \frac{\mathrm{mol}}{\mathrm{s}}\right)(50\ \mathrm{mol}\%)$$

$$= \left(1000\ \frac{\mathrm{mol}}{\mathrm{s}}\right)(0.50)$$

$$= 500\ \mathrm{mol/s\ acetone}$$

The acetone flow in the overhead product (distillate) is that recovered from the feed,

$$Dx_D = Fz_F \times 98\%$$

$$= \left(500\ \frac{\mathrm{mol}}{\mathrm{s}}\right)(0.98)$$

$$= 490\ \mathrm{mol/s\ acetone}$$

The distillate composition is to be 90 mol% acetone, making the mole fraction of acetone in the distillate 0.90. The overhead product (distillate) flow rate is

$$D = \frac{Dx_D}{x_D} = \frac{490\ \dfrac{\mathrm{mol}}{\mathrm{s}}}{90\ \mathrm{mol}\%}$$

$$= \frac{490\ \dfrac{\mathrm{mol}}{\mathrm{s}}}{0.90}$$

$$= 544.44\ \mathrm{mol/s}$$

The total balance around the column is found from

$$F = D + B$$

The bottoms flow rate is then

$$B = F - D$$

$$= 1000\ \frac{\mathrm{mol}}{\mathrm{s}} - 544.44\ \frac{\mathrm{mol}}{\mathrm{s}}$$

$$= 455.56\ \mathrm{mol/s}$$

The acetone balance around the column is found from

$$Fz_F = Dx_D + Bx_B$$

The acetone flow rate in the bottoms is then

$$
\begin{aligned}
Bx_B &= Fz_F - Dx_D \\
&= 500\,\frac{\text{mol}}{\text{s}} - 490\,\frac{\text{mol}}{\text{s}} \\
&= 10\ \text{mol/s}
\end{aligned}
$$

The bottoms acetone composition is

$$
\begin{aligned}
x_B = \frac{Bx_B}{B} &= \frac{10\,\dfrac{\text{mol}}{\text{s}}}{455.56\,\dfrac{\text{mol}}{\text{s}}} \\
&= 0.021\,951 \quad (2.2\ \text{mol\%})
\end{aligned}
$$

The answer is C.

17. Balances around the rectifying section, assuming constant molal overflow, on total moles and acetone are found from

$$V = L + D$$
$$Vy = Lx + Dx_D$$

The rectifying section operating line is then found from the following equation, assuming equal molal overflow.

$$
\begin{aligned}
y &= \frac{L}{V}x + \frac{Dx_D}{V} \\
&= \left(\frac{R_D}{R_D + 1}\right)x + \left(\frac{x_D}{R_D + 1}\right)
\end{aligned}
$$

The operating line with the minimum slope is found from

$$y = \left(\frac{R_{\min}}{R_{\min} + 1}\right)x + \left(\frac{x_D}{R_{\min} + 1}\right)$$

The operating line intersects the $y = x$ line at $x = x_D$. The pinch point was obtained graphically as the line starting at $(0.90, 0.90)$ and drawn tangent to the equilibrium curve. From the following McCabe-Thiele diagram of the minimum reflux ratio operating line pinch, the operating line with minimum slope occurs at the pinch point $(0.60, 0.853)$ with the equilibrium curve.

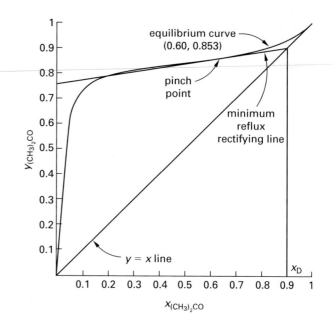

The minimum slope of the operating line is

$$
\begin{aligned}
\left(\frac{L}{V}\right)_{\min} &= \frac{0.90 - 0.853}{0.90 - 0.60} = 0.156\,67 \\
&= \frac{R_{\min}}{R_{\min} + 1}
\end{aligned}
$$

Solving this equation for the minimum external reflux ratio,

$$
\begin{aligned}
R_{\min} &= \frac{\left(\dfrac{L}{V}\right)_{\min}}{1 - \left(\dfrac{L}{V}\right)_{\min}} = \frac{0.156\,67}{1 - 0.156\,67} \\
&= 0.185\,78 \quad (0.18)
\end{aligned}
$$

The answer is A.

18. The actual operating external reflux ratio from the problem statement is eight times the minimum,

$$
\begin{aligned}
R_D = \frac{L}{D} &= 8R_{\min} \\
&= (8)(0.184\,83) \\
&= 1.4786
\end{aligned}
$$

The actual recifying section operating line is plotted in the following McCabe-Thiele diagram.

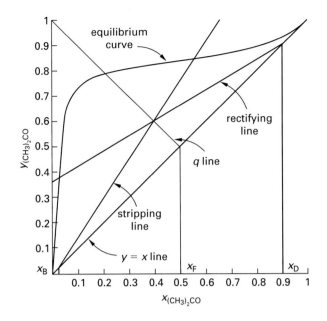

$$y = \left(\frac{R_D}{R_D+1}\right)x + \left(\frac{x_D}{R_D+1}\right)$$
$$= \left(\frac{1.4786}{1.4786+1}\right)x + \left(\frac{0.90}{1.4786+1}\right)$$
$$= 0.597x + 0.363$$

Since the quality of the feed is 50% liquid, the value of q is found from the equation,

$$q \equiv \frac{L_S - L_R}{F} = \frac{L_F}{F} = \frac{0.5F}{F}$$
$$= 0.5$$

The liquid flow in the stripping section is found by solving this equation.

$$L_S = Fq + L_R = Fq + R_D D$$
$$= \left(1000\ \frac{mol}{s}\right)(0.5) + (1.4786)$$
$$\times \left(544.44\ \frac{mol}{s}\right)$$
$$= 1305\ mol/s$$

The feed operating (or q) line is plotted in the McCabe-Thiele diagram.

$$y = \left(\frac{q}{q-1}\right)x - \frac{x_F}{q-1}$$
$$= \left(\frac{0.5}{0.5-1}\right)x - \frac{0.5}{0.5-1}$$
$$= -1x + 1$$

The total and acetone balances around the stripping section again assuming constant molal overflow, are

$$L_S = V_S + B$$
$$L_S x = V_S y + B x_B$$

The vapor flow in the stripping section is

$$V_S = L_S - B$$
$$= 1305\ \frac{mol}{s} - 455.56\ \frac{mol}{s}$$
$$= 849.44\ mol/s$$

The liquid-to-vapor molar flow ratio is

$$\frac{L_S}{V_S} = \frac{1305\ \frac{mol}{s}}{849.44\ \frac{mol}{s}}$$
$$= 1.5363\quad(1.5)$$

The answer is B.

19. Construct the pseudoequilibrium curve using the definition of Murphree plate efficiency, the equilibrium curve, and the operating lines.

$$E_{ME} = \frac{y_n - y_{n+1}}{y_n^* - y_{n+1}}$$
$$= 70\%$$

A pseudoequilibrium curve [8] is drawn between the operating lines and the equilibrium curve by using the fractional vertical distance equal to the Murphree efficiency given. So, in this problem, the curve lies 70% of the distance between the operating lines and the equilibrium curve. The operating lines and pseudoequilibrium curve are shown in the following McCabe-Thiele diagram of actual operation. Valves can be approximated graphically or can be accomplished numerically by using the Murphree plate efficiency calculation and solving for the coordinate of the pseudoequilibrium curve.

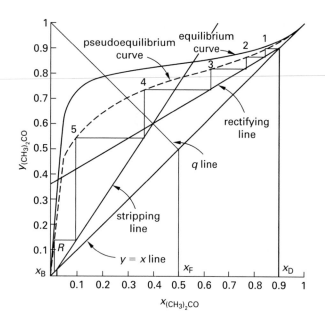

x_n	y_n^*	y_{n+1}	$y_n = y_{n+1} + E_{ME}$ $\times (y_n^* - y_{n+1})$
0	0	0	0
0.1	0.7301	0.142	0.553
0.2	0.7916	0.295	0.643
0.3	0.8124	0.449	0.703
0.4	0.8269	0.602	0.760
0.5	0.8387	0.661	0.786
0.6	0.8532	0.721	0.814
0.7	0.8712	0.781	0.844
0.8	0.8950	0.840	0.879
0.9	0.9335	0.9	0.923
1	1	1	1

The actual stages are stepped off with the result that six stages are required, or five stages and a partial reboiler. A partial reboiler is considered to be an equilibrium stage.

The answer is B.

$$y_n = y_{n+1} + E_{ME} (y_n^* - y_{n+1})$$

The pseudoequilibrium values, y_{n+1}, are obtained for a given liquid composition, x_n, from the operating lines with the equilibrium curve value, y_n^*, of the given liquid composition. The recifying section operating line is plotted in the McCabe-Thiele diagram.

$$y = \left(\frac{R_D}{R_D + 1}\right) x + \left(\frac{x_D}{R_D + 1}\right)$$
$$= \left(\frac{1.4786}{1.4786 + 1}\right) x + \left(\frac{0.90}{1.4786 + 1}\right)$$
$$= 0.597x + 0.363$$

The stripping-section operating line is plotted in the McCabe-Thiele diagram.

$$y = \frac{L_S}{V_S}x - \frac{Bx_B}{V_S}$$
$$= \left(\frac{1305 \, \frac{mol}{s}}{849.44 \, \frac{mol}{s}}\right) x - \frac{10 \, \frac{mol}{s}}{849.44 \, \frac{mol}{s}}$$
$$= 1.54x - 0.0118$$

Some points on the curve are given in the following table. The stripping line is used for the points below the intersection with the q line and the rectifying line is used for the points above the intersection with the q line. This intersection occurs, from the McCabe-Thiele, at about 40 mol% in the liquid.

20. The product benzoic acid molar flow is

$$\left(1000.0 \, \frac{kg \, C_6H_5COOH}{h}\right)\left(\frac{1 \, kmol \, C_6H_5COOH}{122.12 \, kg \, C_6H_5COOH}\right)$$
$$= 8.1887 \, kmol \, C_6H_5COOH/h$$

One mole of toluene is reacted per mole of benzoic acid formed. Toluene reacted is

$$\left(8.1887 \, \frac{kmol \, C_6H_5COOH}{h}\right)\left(1 \, \frac{kmol \, C_6H_5CH_3}{kmol \, C_6H_5COOH}\right)$$
$$= 8.1887 \, kmol \, C_6H_5CH_3/h$$

$$\left(8.1887 \, \frac{kmol \, C_6H_5CH_3}{h}\right)\left(92.13 \, \frac{kg \, C_6H_5CH_3}{kmol \, C_6H_5CH_3}\right)$$
$$= 754.42 \, kg \, C_6H_5CH_3/h$$

Overall yield of toluene is defined as

$$\text{overall yield of toluene} = \frac{\text{toluene reacted}}{\text{fresh toluene in feed}}$$

Solving for fresh toluene in the feed

$$T = \frac{754.42 \, \frac{kg \, C_6H_5CH_3}{h}}{0.90}$$
$$= 838.24 \, kg \, C_6H_5CH_3/h$$

The toluene in product is the fresh toluene in the feed minus the toluene reacted,

$$838.24 \, \frac{kg \, C_6H_5CH_3}{h} - 754.42 \, \frac{kg \, C_6H_5CH_3}{h}$$
$$= 83.82 \, kg \, C_6H_5CH_3/h$$

Total benzoic acid and toluene product flow is

$$P = 1000 \ \frac{\text{kg C}_6\text{H}_5\text{COOH}}{\text{h}} + 83.82 \ \frac{\text{kg C}_6\text{H}_5\text{CH}_3}{\text{h}}$$
$$= 1083.82 \text{ kg/h} \quad (1100 \text{ kg/h})$$

The answer is B.

21. From Prob. 20, the product benzoic acid flow rate is 8.1887 kmol/h. From the reaction, one mole of water is produced per mole of benzoic acid produced. The molar flow of water produced is

$$\left(8.1887 \ \frac{\text{kmol C}_6\text{H}_5\text{COOH}}{\text{h}}\right)\left(1 \ \frac{\text{kmol H}_2\text{O}}{\text{kmol C}_6\text{H}_5\text{COOH}}\right)$$
$$= 8.1887 \text{ kmol H}_2\text{O/h}$$

The wastewater flow rate is

$$W = \left(8.1887 \ \frac{\text{kmol H}_2\text{O}}{\text{h}}\right)\left(18.01 \ \frac{\text{kg H}_2\text{O}}{\text{kmol H}_2\text{O}}\right)$$
$$= 147.48 \text{ kg H}_2\text{O/h} \quad (150 \text{ kg H}_2\text{O/h})$$

The answer is A.

22. From Prob. 20, the product benzoic acid flow rate is 8.1887 kmol/h. From the reaction, $3/2$ mol of O_2 is produced per mole of benzoic acid produced. The molar flow of O_2 reacted is

$$\left(8.1887 \ \frac{\text{kmol C}_6\text{H}_5\text{COOH}}{\text{h}}\right)\left(\frac{\frac{3}{2} \text{ kmol O}_2}{1 \text{ kmol C}_6\text{H}_5\text{COOH}}\right)$$
$$= 12.283 \text{ kmol O}_2/\text{h}$$

The mass flow rate of O_2 reacted is

$$\left(12.283 \ \frac{\text{kmol O}_2}{\text{h}}\right)\left(32.00 \ \frac{\text{kg O}_2}{\text{kmol O}_2}\right)$$
$$= 393.06 \text{ kg O}_2/\text{h}$$

Since 1 kmol of air is composed of approximately 0.79 kmol of nitrogen and 0.21 kmol of oxygen, the molar flow rate of nitrogen coming in with the air is

$$\left(12.283 \ \frac{\text{kmol O}_2}{\text{h}}\right)\left(\frac{0.79 \text{ kmol N}_2}{0.21 \text{ kmol O}_2}\right)$$
$$= 46.207 \text{ kmol N}_2/\text{h}$$

The mass flow rate of nitrogen coming in with the air is

$$N_2 = \left(46.207 \ \frac{\text{kmol N}_2}{\text{h}}\right)\left(28.02 \ \frac{\text{kg N}_2}{\text{kmol N}_2}\right)$$
$$= 1294.7 \text{ kg N}_2/\text{h}$$

The total mass flow rate of air is

$$A = 393.06 \ \frac{\text{kg O}_2}{\text{h}} + 1294.7 \ \frac{\text{kg N}_2}{\text{h}}$$
$$= 1687.8 \text{ kg air/h} \quad (1700 \text{ kg air/h})$$

The answer is B.

23. From Prob. 22, the molar flow rate of air is

$$12.283 \ \frac{\text{kmol O}_2}{\text{h}} + 46.207 \ \frac{\text{kmol N}_2}{\text{h}}$$
$$= 58.49 \text{ kmol air/h}$$

The volumetric flow rate of air from the ideal gas law is

$$Q = \frac{F_{\text{air}}RT}{P}$$
$$= \frac{\left(58.49 \ \frac{\text{kmol}}{\text{h}}\right)\left(8.314510 \ \frac{\text{kPa·m}^3}{\text{kmol·K}}\right)}{(1 \text{ atm})\left(101.33 \ \frac{\text{kPa}}{\text{atm}}\right)}$$
$$= 1310.9 \text{ m}^3/\text{h} \quad (1300 \text{ m}^3/\text{h})$$

The answer is D.

24. For a flow reactor at steady state, the conversion of toluene, X_T, is defined as [5]

$$X_T = \frac{F_{To} - F_T}{F_{To}}$$

F_{To} is the toluene entering the reactor and F_T is the toluene leaving the reactor.

Consider the following simplified diagram of the toluene flows (from Prob. 20).

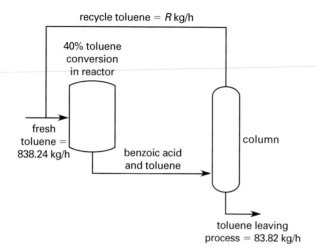

recycle toluene = R kg/h

40% toluene conversion in reactor

fresh toluene = 838.24 kg/h

benzoic acid and toluene

column

toluene leaving process = 83.82 kg/h

The toluene entering the reactor is the fresh toluene in the feed plus the unknown recycle toluene,

$$F_T = 838.24 \, \frac{\text{kg } C_6H_5CH_3}{\text{h}} + R$$

The toluene leaving the reactor is the toluene leaving the process plus the unknown recycle toluene,

$$F_T = 83.82 \, \frac{\text{kg } C_6H_5CH_3}{\text{h}} + R$$

The conversion of toluene equation becomes

$$X_T = \frac{F_{To} - F_T}{F_{To}}$$

$$0.40 = \frac{\left(83.82 \, \frac{\text{kg } C_6H_5CH_3}{\text{h}} + R\right) - \left(838.24 \, \frac{\text{kg } C_6H_5CH_3}{\text{h}} + R\right)}{R + 83.82 \, \frac{\text{kg } C_6H_5CH_3}{\text{h}}}$$

Rearranging and solving for recycle flow,

$$R = \left(1047.8 \, \frac{\text{kg}}{\text{h}}\right) \left(1000 \, \frac{\text{kg}}{\text{h}}\right)$$

The answer is A.

25. The steady-state biomass overall balance is

$$0 = F_1 X_1 - F_4 X_4 - F_6 X_6 + r_X V$$

F_1 volumetric flow rate
      in stream 1   L/d
X_1 biomass concentration
      in stream 1   mg biomass/L

r_X rate of biomass production   mg biomass/d·L
V volume of the bioreactor   L

Substituting the rate expression for biomass into the biomass overall balance,

$$X_1 = 0$$

$$F_4 X_4 + F_6 X_6 = \left(\frac{\mu_{\max} S}{K_S + S} - k_e\right) X_3 V$$

Rearranging to get sludge age [1],

$$\theta_X = \frac{X_3 V}{F_4 X_4 + F_6 X_6}$$

$$= \left(\frac{\mu_{\max} S}{K_S + S} - k_e\right)^{-1}$$

$$= \left(\frac{\mu_{\max} S}{K_S + S} - k_e\right)^{-1}$$

$$= \left(\frac{(2.5 \text{ d}^{-1}) \left(10 \, \frac{\text{mg}}{\text{L}}\right)}{50 \, \frac{\text{mg}}{\text{L}} + 10 \, \frac{\text{mg}}{\text{L}}} - 0.05 \text{ d}^{-1}\right)^{-1}$$

$$= 2.7273 \text{ d} \quad (3 \text{ d})$$

The answer is B.

26. There is no biomass in stream $1 (X_1 = 0)$. A total biomass balance, assuming constant density, and a biomass balance around the bioreactor and recycle addition point gives the flow of stream 1 plus stream 7 equals stream 3.

$$F_1 + F_7 = F_3$$

$$0 = F_7 X_7 - F_3 X_3 + \left(\frac{\mu_{\max} S}{K_S + S} - k_e\right) X_3 V$$

The recycle ratio, R, is defined as the ratio of the flow of stream 7 to stream 1.

$$R = \frac{F_7}{F_1}$$

$$F_7 = F_1 R$$

The reactor effluent flow is

$$F_3 = F_1 + F_7 = F_1 + R F_1$$
$$= F_1 (1 + R)$$

Substituting this equation into the total biomass balance,

$$0 = F_7 X_7 - F_1 (1 + R) X_3 + \left(\frac{\mu_{\max} S}{K_S + S} - k_e\right) X_3 V$$

Substituting the value of the sludge age into this equation,

$$0 = F_1 R X_7 - F_1(1+R)X_3 + \frac{1}{\theta_X}X_3 V$$

Solving for the volume of the bioreactor [1],

$$V = \theta_X F_1 \left(1 + R - R\frac{X_7}{X_3}\right)$$

The problem statement gives the biomass sludge concentration in the recycle as 1.3 times the biomass concentration of the flow leaving the reactor or $X_7 = 1.3X_3$. The recycle ratio is $R = 3.2$. Inserting these numerical values, the volume of the bioreactor is,

$$V = \theta_X F_1 \left(1 + R - R\frac{X_7}{X_3}\right)$$
$$= (2.7273 \text{ d}) \left(10\,000\ \frac{\text{L}}{\text{d}}\right)(1 + 3.2 - (3.2)(1.3))$$
$$= 1090.9 \text{ L} \quad (1100 \text{ L})$$

The answer is D.

27. A substrate balance around the bioreactor and recycle addition point is

$$0 = F_1 S_1 + F_7 S_7 - F_3 S_3 - r_S V$$

Note that the problem statement says that the substrate in stream 3 is the same as in stream 7.

$$S_3 = S_7$$

After some algebra and inserting the expression for sludge age and the expressions for recycle ratio, the substrate balance gives the biomass concentration in the bioreactor [8].

$$r_S V = F_1 S_1 + F_7 S_7 - F_3 S_3$$
$$\left(\frac{\mu_{\max} S}{K_S + S}\right)\left(\frac{X_3}{Y_{X/S}}\right)V = F_1 S_1 + R F_1 S_3$$
$$- F_1(1+R)S_3$$
$$\left(\frac{1}{\theta_X} + k_e\right)\left(\frac{X_3}{Y_{X/S}}\right)V = F_1 S_1 + R F_1 S_3$$
$$- F_1 S_3 - F_1 R S_3$$
$$(1 + \theta_X k_e)\left(\frac{X_3}{Y_{X/S}}\right)V = \theta_X (F_1 S_1 - F_1 S_3)$$

$$X_3 = \frac{Y_{X/S}\theta_X F_1 (S_1 - S_3)}{V(1 + \theta_X k_e)}$$

Inserting numerical values, the biomass concentration in the bioreactor is

$$X_3 = \frac{\left(0.5\ \frac{\text{mg biomass}}{\text{mg substrate}}\right)(2.7273 \text{ d})\left(10\,000\ \frac{\text{m}^3}{\text{d}}\right)}{(1090.9 \text{ L})(1 + (2.7273 \text{ d})(0.05\ d^{-1}))}$$
$$\times \left(500\ \frac{\text{mg substrate}}{\text{L}} - 10\ \frac{\text{mg substrate}}{\text{L}}\right)$$
$$= 5390 \text{ mg biomass/L} \quad (5400 \text{ mg biomass/L})$$

The answer is D.

28. First, the system must be linearized about the steady-state point. Then, from the NCEES Handbook, Laplace transforms can be taken to yield the final transfer function [3,9]. The problem statement says that only C_A, T, and T_j change. A Taylor series for these three variables about the steady-state point for the mass and energy balances is

$$\frac{dC_A}{dt} = f_1(C_A, T, T_j)$$
$$\approx f_1(C_{As}, T_s, T_{js}) + \left(\frac{\partial f_1}{\partial C_A}\right)_s (C_A - C_{As})$$
$$+ \left(\frac{\partial f_1}{\partial T}\right)_s (T - T_s) + \left(\frac{\partial f_1}{\partial T_j}\right)_s (T_j - T_{js})$$
$$\frac{dT}{dt} = f_2(C_A, T, T_j)$$
$$\approx f_2(C_{As}, T_s, T_{js}) + \left(\frac{\partial f_2}{\partial C_A}\right)_s (C_A - C_{As})$$
$$+ \left(\frac{\partial f_2}{\partial T}\right)_s (T - T_s) + \left(\frac{\partial f_2}{\partial T_j}\right)_s (T_j - T_{js})$$

At the steady state point,

$$\frac{dC_{As}}{dt} = f_1(C_{As}, T_s, T_{js})$$
$$= 0$$
$$\frac{dT_s}{dt} = f_2(C_{As}, T_s, T_{js})$$
$$= 0$$

The deviation variables about the steady-state point are defined as

$$y_1 = C_A - C_{As}$$
$$y_2 = T - T_s$$
$$x_1 = T_j - T_{js}$$

When the steady-state equations are subtracted from the unsteady-state equations and the deviation variable is substituted, the balances become

$$\frac{dy_1}{dt} = \left(\frac{\partial f_1}{\partial C_A}\right)_s y_1 + \left(\frac{\partial f_1}{\partial T}\right)_s y_2 + \left(\frac{\partial f_1}{\partial T_j}\right)_s x_1$$

$$\frac{dy_2}{dt} = \left(\frac{\partial f_2}{\partial C_A}\right)_s y_1 + \left(\frac{\partial f_2}{\partial T}\right)_s y_2 + \left(\frac{\partial f_2}{\partial T_j}\right)_s x_1$$

$$y_1(0) = y_2(0)$$

$$= 0$$

The system can be rewritten with constants for the partial derivatives as in the Control section of the NCEES Handbook.

$$\frac{dy_1}{dt} = a_{11}y_1 + a_{12}y_2 + b_{11}x_1$$

$$\frac{dy_2}{dt} = a_{21}y_1 + a_{22}y_2 + b_{21}x_1$$

$$y_1(0) = y_2(0)$$

$$= 0$$

Take Laplace transforms of this system as in the Mathematics section of the NCEES Handbook.

$$sY_1(s) - y_1(0) = a_{11}Y_1(s) + a_{12}Y_2(s) + b_{11}X_1(s)$$

$$sY_2(s) - y_2(0) = a_{21}Y_1(s) + a_{22}Y_2(s) + b_{21}X_1(s)$$

Since the initial conditions are zero, this becomes

$$sY_1(s) = a_{11}Y_1(s) + a_{12}Y_2(s) + b_{11}X_1(s)$$

$$sY_2(s) = a_{21}Y_1(s) + a_{22}Y_2(s) + b_{21}X_1(s)$$

Now the transfer function must be obtained by solving the system,

$$(s - a_{11})Y_1(s) - a_{12}Y_2(s) = b_{11}X_1(s)$$

$$-a_{21}Y_1(s) + (s - a_{22})Y_2(s) = b_{21}X_1(s)$$

Or in matrix notation,

$$\begin{bmatrix} s - a_{11} & -a_{12} \\ -a_{21} & s - a_{22} \end{bmatrix} \begin{bmatrix} Y_1(s) \\ Y_2(s) \end{bmatrix} = \begin{bmatrix} b_{11}X_1(s) \\ b_{21}X_1(s) \end{bmatrix}$$

$$\begin{bmatrix} Y_1(s) \\ Y_2(s) \end{bmatrix} = \begin{bmatrix} s - a_{11} & -a_{12} \\ -a_{21} & s - a_{22} \end{bmatrix}^{-1} \begin{bmatrix} b_{11}X_1(s) \\ b_{21}X_1(s) \end{bmatrix}$$

Cramer's rule is used to solve the system of equations,

$$Y_1(s) = \frac{\begin{vmatrix} b_{11}X_1(s) & -a_{12} \\ b_{21}X_1(s) & s - a_{22} \end{vmatrix}}{\begin{vmatrix} s - a_{11} & -a_{12} \\ -a_{21} & s - a_{22} \end{vmatrix}}$$

Expanding the determinants, the transfer function, as in the Control section of the NCEES Handbook, is

$$Y_1(s) = \frac{b_{11}X_1(s)(s - a_{22}) - (-a_{12})(b_{21}X_1(s))}{(s - a_{11})(s - a_{22}) - (-a_{12})(-a_{21})}$$

$$= \left(\frac{(s - a_{22})b_{11} + a_{12}b_{21}}{s^2 - (a_{11} + a_{22})s + a_{11}a_{22} - a_{12}a_{21}}\right) X_1(s)$$

The coefficients needed are obtained from the six partial derivatives

$$a_{11} = \left(\frac{\partial f_1}{\partial C_A}\right)_s$$

$$= \left(\frac{\partial}{\partial C_A}\left(\frac{F}{V}(C_{Af} - C_A) - k_o e^{\frac{-E}{RT}} C_A\right)\right)_s$$

$$= -\frac{F}{V} - k_o e^{\frac{-E}{RT_s}}$$

$$= -\frac{0.1\,\frac{m^3}{s}}{1\,m^3} - \left(7.295 \times 10^5\,s^{-1}\right)$$

$$\times e^{\frac{-8.3736 \times 10^7\,\frac{J}{kmol}}{\left(8314.510\,\frac{J}{kmol \cdot K}\right)(640.51K)}}$$

$$= -0.208\,24\,s^{-1}$$

$$a_{12} = \left(\frac{\partial f_1}{\partial T}\right)_s$$

$$= \left(\frac{\partial}{\partial T}\left(\frac{F}{V}(C_{Af} - C_A) - k_o e^{\frac{-E}{RT}} C_A\right)\right)_s$$

$$= -\frac{E}{RT_s^2} k_o e^{\frac{-E}{RT_s}} C_{As}$$

$$= -\frac{8.3736 \times 10^7\,\frac{J}{kmol}}{\left(8314.510\,\frac{J}{kmol \cdot K}\right)(640.51K)^2}$$

$$\times (7.295 \times 10^5\,s^{-1})\, e^{\frac{-8.3736 \times 10^7\,\frac{J}{kmol}}{\left(8314.510\,\frac{J}{kmol \cdot K}\right)(640.51K)}}$$

$$\times \left(4.8020\,\frac{kmol}{m^3}\right)$$

$$= -1.2760 \times 10^{-2}\,kmol/m^3 \cdot s \cdot K$$

$$b_{11} = \left(\frac{\partial f_1}{\partial T_j}\right)_s$$

$$= \left(\frac{\partial}{\partial T_j}\left(\frac{F}{V}(C_{Af} - C_A) - k_o e^{\frac{-E}{RT}} C_A\right)\right)_s$$

$$= 0$$

$$a_{21} = \left(\frac{\partial f_2}{\partial C_A}\right)_s$$

$$= \left(\frac{\partial}{\partial C_A}\left(\begin{array}{c}\frac{F}{V}(T_f - T) + \frac{-\Delta H}{\rho c_p}k_o e^{\frac{-E}{RT}}\\ \times C_A - \frac{UA}{V\rho c_p}(T - T_j)\end{array}\right)\right)_s$$

$$= \frac{-\Delta H}{\rho c_p}k_o e^{\frac{-E}{RT_s}}$$

$$= \frac{\left(1.6747 \times 10^8 \frac{J}{kmol}\right)}{\left(1000 \frac{kg}{m^3}\right)\left(5000 \frac{J}{kg \cdot K}\right)}(7.295 \times 10^5 \text{ s}^{-1})$$

$$\times e^{\frac{-8.3736 \times 10^7 \frac{J}{kmol}}{\left(8314.510 \frac{J}{kmol \cdot K}\right)(640.51K)}}$$

$$= 3.6254 \text{ m}^3 \cdot K/s \cdot kmol$$

$$b_{21} = \left(\frac{\partial f_2}{\partial T_j}\right)_s$$

$$= \left(\frac{\partial}{\partial T_j}\left(\begin{array}{c}\frac{F}{V}(T_f - T) + \frac{-\Delta H}{\rho c_p}k_o e^{\frac{-E}{RT}}\\ \times C_A - \frac{UA}{V\rho c_p}(T - T_j)\end{array}\right)\right)_s$$

$$= \frac{UA}{V\rho c_p}$$

$$= \frac{\left(1.0425 \times 10^5 \frac{J}{s \cdot m^2 \cdot K}\right)(1 \text{ m}^2)}{(1 \text{ m}^3)\left(1000 \frac{kg}{m^3}\right)\left(5000 \frac{J}{kg \cdot K}\right)}$$

$$= 0.02085 \text{ s}^{-1}$$

$$a_{22} = \left(\frac{\partial f_2}{\partial T}\right)_s$$

$$= \left(\frac{\partial}{\partial T}\left(\begin{array}{c}\frac{F}{V}(T_f - T) + \frac{-\Delta H}{\rho c_p}k_o e^{\frac{-E}{RT}}\\ \times C_A - \frac{UA}{V\rho c_p}(T - T_j)\end{array}\right)\right)_s$$

$$= -\frac{F}{V} + \left(\frac{-\Delta H}{\rho c_p}\right)\left(\frac{E}{RT_s^2}\right)k_o e^{\frac{-E}{RT}}C_{As} - \frac{UA}{V\rho c_p}$$

$$= -\frac{F}{V} + \left(\frac{-\Delta H}{\rho c_p}\right)(-a_{12}) - b_{21}$$

$$= -\frac{0.1 \frac{m^3}{s}}{1 \text{ m}^3} + \left(\frac{1.6747 \times 10^8 \frac{J}{kmol}}{\left(1000 \frac{kg}{m^3}\right)\left(5000 \frac{J}{kg \cdot K}\right)}\right)$$

$$\times \left(1.2760 \times 10^{-2} \frac{kmol}{m^3 \cdot s \cdot K}\right) - 0.02085 \text{ s}^{-1}$$

$$= 0.30652 \text{ s}^{-1}$$

The transfer function is then

$$Y_1(s) = \left(\frac{(s - a_{22})b_{11} + a_{12}b_{21}}{s^2 - (a_{11} + a_{22})s + a_{11}a_{22} - a_{12}a_{21}}\right)X_1(s)$$

$$= \frac{(s - a_{22})(0) + \left(-1.2760 \times 10^{-2} \frac{kmol}{m^3 \cdot s \cdot K}\right)}{\begin{array}{c}\times (0.02085 \text{ s}^{-1})\\ s^2 - (-0.20824 \text{ s}^{-1} + 0.30652 \text{ s}^{-1})s\\ + (-0.20824 \text{ s}^{-1})(0.30652 \text{ s}^{-1})\\ - \left(-1.2760 \times 10^{-2} \frac{kmol}{m^3 \cdot s \cdot K}\right)\\ \times \left(3.6254 \frac{m^3 \cdot K}{s \cdot kmol}\right)\end{array}}X_1(s)$$

$$Y_1(s)\left[\frac{kmol}{m^3}\right] = \left(\frac{-2.6605 \times 10^{-4}}{s^2 - 0.09828s - 0.017570}\right)X_1(s)\text{ [K]}$$

Rewritten in terms of the physical variables, the transfer function is

$$\frac{C_A(s)\left[\frac{kmol}{m^3}\right]}{T_j(s)\text{ [K]}} = \frac{-2.7 \times 10^{-4}}{s^2 - 0.098s - 0.018}$$

The answer is B.

29. The open-loop characteristic equation from the Control section of the NCEES Handbook is

$$s^2 - 0.09828s - 0.017570 = 0$$

The first root of the open-loop characteristic equation is

$$s = \frac{-0.09828 + \sqrt{\begin{array}{c}(-0.09828)^2 - (4)\\ \times (1)(-0.017570)\end{array}}}{(2)(1)}\text{ s}^{-1}$$

The second root of the open-loop characteristic equation is

$$s = \frac{-0.09828 - \sqrt{\begin{array}{c}(-0.09828)^2 - (4)\\ \times (1)(-0.017570)\end{array}}}{(2)(1)}\text{ s}^{-1}$$

Since one of the roots is a positive real number, the point is unstable.

The answer is B.

30. The proportional plus derivative controller equation from the Control section of the NCEES Handbook is

$$G_c(s) = K(1 + T_D s)$$

The open-loop transfer function from the Control section of the NCEES Handbook is

$$G_1(s) = \frac{Y_1(s)}{X_1(s)} = \frac{-2.6605 \times 10^{-4}}{s^2 - 0.098\,28s - 0.017\,570} \left[\frac{\text{kmol}}{\text{m}^3\,\text{K}}\right]$$

The closed-loop transfer function from the Control section in the NCEES Handbook is

$$\frac{C(s)}{R(s)} = \frac{G_c(s)G_1(s)}{1 + G_c(s)G_1(s)}$$

$$\frac{C(s)\left[\frac{\text{kmol}}{\text{m}^3}\right]}{R(s)\left[\frac{\text{kmol}}{\text{m}^3}\right]} = \frac{K\left[\frac{\text{m}^3\cdot\text{K}}{\text{kmol}}\right](1 + T_D\,[\text{s}]\,s)}{1 + K\left[\frac{\text{m}^3\cdot\text{K}}{\text{kmol}}\right](1 + T_D\,[\text{s}]\,s)} \times \frac{\left(\frac{-2.6605 \times 10^{-4}}{s^2 - 0.098\,28s - 0.017\,570}\right)}{\left(\frac{-2.6605 \times 10^{-4}}{s^2 - 0.098\,28s - 0.017\,570}\right)} \times \left[\frac{\text{kmol}}{\text{m}^3\cdot\text{K}}\right]$$

The closed-loop characteristic equation is

$$1 + K(1 + T_Ds)\left(\frac{-2.6605 \times 10^{-4}}{s^2 - 0.098\,28s - 0.017\,570}\right) = 0$$
$$s^2 + \left(-0.098\,28 - 2.6605 \times 10^{-4}KT_D\right)s + \left(-0.017\,570 - (2.6605 \times 10^{-4})K\right) = 0$$

For a second-order characteristic equation,

$$a_os^2 + a_1s + a_2 = 0$$

The Roth array is from the Control section of the NCEES Handbook.

$$\begin{array}{cc} a_o & a_2 \\ a_1 & a_3 = 0 \\ b_1 \end{array}$$

$$b_1 = \frac{a_1a_2 - a_0a_3}{a_1} = a_2$$

The Roth array for this system becomes

$$\begin{array}{cc} 1 & -0.017\,570 \\ & -(2.6605 \times 10^{-4})K \\ -0.098\,28 & \\ -(2.6605 \times 10^{-4})KT_D & 0 \\ -0.017\,570 & \\ -(2.6605 \times 10^{-4})K \end{array}$$

All items in the first column must be positive for stability.

$$-0.098\,28 - (2.6605 \times 10^{-4})KT_D > 0$$
$$-0.017\,570 - (2.6605 \times 10^{-4})K > 0$$

The first element will be positive if

$$-0.098\,28 - (2.6605 \times 10^{-4})KT_D > 0$$
$$KT_D < -370 \text{ m}^3\cdot\text{K}\cdot\text{s/kmol}$$

The second element will be positive if

$$-0.0175\,70 - (2.6605 \times 10^{-4})K > 0$$
$$K < -66 \text{ m}^3\cdot\text{K/kmol}$$

The system is stable for all $K < -66$ m$^3\cdot$K/kmol and $KT_D < -370$ m$^3\cdot$K$\cdot$s/kmol.

The answer is D.

31. The pipe length, inside insulation diameter, inside insulation temperature (assuming no resistance through the pipe wall as given in the problem statement), and ambient temperature are

$$L = 1 \text{ m}$$
$$D_i = 0.08 \text{ m}$$
$$T_i = 650\text{K}$$
$$T_a = 300\text{K}$$

The insulation thickness is 0.05 m, so the insulation outer diameter is

$$t = 0.05 \text{ m}$$
$$D_o = D_i + 2t = 0.08 \text{ m} + (2)(0.05 \text{ m})$$
$$= 0.18 \text{ m}$$

Initially, guess that the outer surface temperature of the pipe is

$$T_o = 350\text{K}$$

This value will need to be checked and corrected if necessary. The Stefan-Boltzman constant [3] is

$$\sigma = 5.67 \times 10^{-8} \text{ W/m}^2\cdot\text{K}^4$$

The radiation heat-transfer coefficient from the Heat Transfer section of the NCEES Handbook is

$$h_r = \frac{\varepsilon\sigma\left(T_o^4 - T_a^4\right)}{T_o - T_a}$$
$$= \frac{(0.05)\left(5.67 \times 10^{-8}\,\frac{\text{W}}{\text{m}^2\cdot\text{K}^4}\right) \times ((350\text{K})^4 - (300\text{K})^4)}{350\text{K} - 300\text{K}}$$
$$= 0.391\,58 \text{ W/m}^2\cdot\text{K}$$

The Prandtl number from the Heat Transfer section of the NCEES Handbook is

$$\Pr = \frac{c_p \mu}{k}$$

$$= \frac{\left(1007 \ \frac{J}{kg \cdot K}\right)\left(1.95 \times 10^{-5} \ \frac{kg}{s \cdot m}\right)}{0.0282 \ \frac{W}{m \cdot K}}$$

$$= 0.696\,33$$

The kinematic viscosity from the Heat Transfer section in the NCEES Handbook is

$$\nu = \frac{\mu}{\rho} = \frac{1.95 \times 10^{-5} \ \frac{kg}{s \cdot m}}{1.088 \ \frac{kg}{m^3}}$$

$$= 1.7923 \times 10^{-5} \, \text{m}^2/\text{s}$$

The coefficient of thermal expansion from the Heat Transfer section of the NCEES Handbook is

$$\beta = \frac{2}{T_0 + T_a} = \frac{2}{350K + 300K}$$

$$= 3.0769 \times 10^{-3} \text{K}^{-1}$$

The acceleration of gravity [3] is

$$g = 9.807 \ \text{m/s}^2$$

For free convection, the long horizontal cylinder Rayleigh number from the Heat Transfer section in the NCEES Handbook is

$$Ra_D = \left(\frac{g\beta(T_s - T_\infty)D^3}{\nu^2}\right)\Pr$$

$$= \left(\frac{g\beta(T_o - T_a)D_o^3}{\nu^2}\right)\Pr$$

$$= \frac{\left(9.807 \ \frac{m}{s^2}\right)(3.0769 \times 10^{-3}\text{K}^{-1})}{\left(1.7923 \times 10^{-5} \ \frac{m^2}{s}\right)^2}$$

$$= 1.9074 \times 10^7$$

The outer convective heat-transfer coefficient is

$$h_o = C\left(\frac{k}{D_o}\right)(Ra_D)^n$$

$$= (0.125)\left(\frac{0.0282 \ \frac{W}{m \cdot K}}{0.18 \ m}\right)(1.9074 \times 10^7)^{0.333}$$

$$= 5.2032 \ \text{W/m}^2 \cdot \text{K}$$

The total convective and radiation heat transfer coefficient is

$$h_r + h_o = 0.391\,58 \ \frac{W}{m^2 \cdot K} + 5.2032 \ \frac{W}{m^2 \cdot K}$$

$$= 5.5948 \ \text{W/m}^2 \cdot \text{K} \quad (5.6 \ \text{W/m}^2 \cdot \text{K})$$

The answer is A.

32. The heat loss, neglecting pipe inner heat-transfer coefficient and pipe wall conduction, from the Heat Transfer section of the NCEES Handbook is

$$\dot{Q} = \frac{\pi D_o L (T_i - T_a)}{\frac{D_o}{2k}\ln\frac{D_o}{D_i} + \frac{1}{h_r + h_o}}$$

$$= \frac{\pi (0.18 \ m)(1 \ m)(650K - 300K)}{\left(\frac{0.18 \ m}{(2)\left(0.073 \ \frac{W}{m \cdot K}\right)}\right)\ln\left(\frac{0.18 \ m}{0.08 \ m}\right) + \frac{1}{5.5948 \ \frac{W}{m^2 \cdot K}}}$$

$$= 167.94 \ \text{W}$$

Check the assumption of the outer surface temperature. The heat loss through the insulation, from the Heat Transfer section in the NCEES Handbook, is

$$\dot{Q} = \frac{\pi D_o L (T_i - T_o)}{\frac{D_o}{2k}\ln\frac{D_o}{D_i}}$$

Solving this equation for outer surface temperature,

$$T_o = T_i - \frac{\dot{Q}}{2\pi k L}\ln\frac{D_o}{D_i}$$

$$= 650K - \left(\frac{167.94 \ W}{2\pi\left(0.073 \ \frac{W}{m \cdot K}\right)(1 \ m)}\right)\ln\left(\frac{0.18 \ m}{0.08 \ m}\right)$$

$$= 353.08K$$

This is close enough to the assumed outer surface temperature of 350K so that no further iteration is required.

The annual energy cost is

$$A_E = (167.94 \ \text{W})\left(8760 \ \frac{h}{yr}\right)\left(3600 \ \frac{s}{h}\right)$$

$$\times \left(\frac{\$10}{1 \ GJ}\right)\left(10^{-9} \ \frac{GJ}{J}\right)$$

$$= \$52.962/\text{yr} \quad (\$53/\text{yr})$$

The answer is D.

33. Evaluate the EUAC of each of the cases, beginning with the thinnest insulation case A. The case with the smallest EUAC is the optimal case. Initial installation cost of the 0.05 m thick insulation is

$$\left[\frac{\$}{\text{m}} \right] = 2340.4 \left(D_i \, [\text{m}] \right)^{0.6} \left(t \, [\text{m}] \right)$$

$$P_I = (2340.4)(0.08 \text{ m})^{0.6} (0.05 \text{ m})(1 \text{ m})$$

$$= \$25.711$$

The equivalent uniform annual cost from the Engineering Economics section of the NCEES Handbook is the annual energy cost plus the annualized initial installation cost of the insulation.

$$\text{EUAC}_A = \frac{\$52.962}{1 \text{ yr}} + 25.711 \left(A/P, 50\%, 10 \right) \frac{\$}{\text{yr}}$$

$$= \frac{\$52.962}{1 \text{ yr}} + (25.711)$$

$$\times \left(\frac{(0.5)(1 + 0.5)^{10}}{(1 + 0.5)^{10} - 1} \right) \frac{\$}{\text{yr}}$$

$$= \$66.044/\text{yr}$$

Now calculate the costs for 0.08 m insulation,

$$t = 0.08 \text{ m}$$

$$D_o = D_i + 2t$$

$$= 0.08 \text{ m} + (2)(0.08 \text{ m})$$

$$= 0.24 \text{ m}$$

The heat loss through the insulation and outer film, including radiation, from the Heat Transfer section of the NCEES Handbook is

$$\dot{Q} = \frac{\pi D_o L \left(T_i - T_a \right)}{\dfrac{D_o}{2k} \ln \dfrac{D_o}{D_i} + \dfrac{1}{h_r + h_o}}$$

$$= \frac{\pi (0.24 \text{ m})(1 \text{ m})(650\text{K} - 300\text{K})}{\left(\dfrac{0.24 \text{ m}}{(2)\left(0.073 \, \dfrac{\text{W}}{\text{m·K}} \right)} \right) \ln \left(\dfrac{0.24 \text{ m}}{0.08 \text{ m}} \right) + \dfrac{1}{5.5948 \, \dfrac{\text{W}}{\text{m}^2 \text{·K}}}}$$

$$= 132.97 \text{ W}$$

The annual energy cost is

$$A_E = (132.97 \text{ W}) \left(8760 \, \frac{\text{h}}{\text{yr}} \right) \left(3600 \, \frac{\text{s}}{\text{h}} \right)$$

$$\times \left(\frac{\$10}{1 \text{ GJ}} \right) \left(10^{-9} \frac{\text{GJ}}{\text{J}} \right)$$

$$= \$41.933/\text{yr}$$

The initial insulation cost is

$$P_I = (2340.4)(0.08 \text{ m})^{0.6}(0.08 \text{ m})(1 \text{ m})$$

$$= \$41.137$$

The equivalent uniform annual cost from the Economics section of the NCEES Handbook is

$$\text{EUAC}_B = A_E + P_I \left(A/P, 50\%, 10 \right)$$

$$= \frac{\$41.933}{1 \text{ yr}} + \frac{41.137 \left(A/P, 50\%, 10 \right)}{\dfrac{\$}{\text{yr}}}$$

$$= \frac{\$41.933}{1 \text{ yr}} + \frac{41.137 \left(\dfrac{(0.5)(1 + 0.5)^{10}}{(1 + 0.5)^{10} - 1} \right)}{\dfrac{\$}{\text{yr}}}$$

$$= \$62.864/\text{yr}$$

Since $\text{EUAC}_A > \text{EUAC}_B$, calculate the costs for 0.11 m insulation thickness.

$$t = 0.11 \text{ m}$$

$$D_o = D_i + 2t$$

$$= 0.08 \text{ m} + (2)(0.11 \text{ m})$$

$$= 0.3 \text{ m}$$

The heat loss is

$$\dot{Q} = \frac{\pi D_o L \left(T_i - T_a \right)}{\dfrac{D_o}{2k} \ln \dfrac{D_o}{D_i} + \dfrac{1}{h_r + h_o}}$$

$$= \frac{\pi (0.3 \text{ m})(1 \text{ m})(650\text{K} - 300\text{K})}{\left(\dfrac{0.3 \text{ m}}{(2)\left(0.073 \, \dfrac{\text{W}}{\text{m·K}} \right)} \right) \ln \left(\dfrac{0.3 \text{ m}}{0.08 \text{ m}} \right) + \dfrac{1}{5.5948 \, \dfrac{\text{W}}{\text{m}^2 \text{·K}}}}$$

$$= 113.96 \text{ W}$$

Annual energy cost is

$$A_E = (113.96 \text{ W}) \left(8760 \, \frac{\text{h}}{\text{yr}} \right) \left(3600 \, \frac{\text{s}}{\text{h}} \right)$$

$$\times \left(\frac{\$10}{1 \text{ GJ}} \right) \left(10^{-9} \frac{\text{GJ}}{\text{J}} \right)$$

$$= \$35.938/\text{yr}$$

Initial insulation cost is

$$P_I = (2340.4)(0.08 \text{ m})^{0.6}(0.11 \text{ m})(1 \text{ m})$$

$$= \$56.564$$

The equivalent uniform annual cost is

$$\text{EUAC}_C = \frac{\$35.938}{1 \text{ yr}} + (56.564)\left(\frac{(0.5)(1+0.5)^{10}}{(1+0.5)^{10}-1}\right)\frac{\$}{\text{yr}}$$

$$= \$64.719/\text{yr}$$

Since $\text{EUAC}_A > \text{EUAC}_B < \text{EUAC}_C$, 0.08 m is the optimum thickness.

The answer is B.

There is no need to check case D since it will have an even larger equivalent uniform annual cost.

34. The production rate of B is

$$R_B = 1\,000\,000 \frac{\text{mol}}{\text{yr}}$$

$$= \left(1\,000\,000 \frac{\text{mol}}{\text{yr}}\right)\left(\frac{1 \text{ yr}}{8760 \text{ h}}\right)\left(\frac{1 \text{ h}}{3600 \text{ s}}\right)$$

$$= 0.031\,710 \text{ mol/s}$$

Conversion for a CSTR is

$$X_A = \frac{F_{Ao} - F_A}{F_{Ao}}$$

The production rate in terms of conversion is

$$R_B = F_{Ao} - F_A = F_{Ao}X_A$$

The unit cost of reactant A is

$$U = \$0.10/\text{mol}$$

The reactor investment per unit volume is

$$P = \$100,000/\text{m}^3$$

For a constant-volume CSTR, the material balance is Eq. 2.32

$$\frac{\tau}{C_{Ao}} = \frac{V}{F_{Ao}} = \frac{X_A}{-r_A}$$

For second-order kinetics,

$$-r_A = kC_A^2 = k\left[C_{Ao}(1 - X_A)\right]^2$$

The feed concentration is

$$C_{Ao} = 1000 \text{ mol/m}^3$$

The rate coefficient is

$$k = 0.000\,01 \text{ m}^3/\text{s·mol}$$

The volume of the reactor is

$$V = \frac{F_{Ao}X_A}{-r_A} = \frac{F_{Ao}X_A}{k\left(C_{Ao}(1 - X_A)\right)^2}$$

$$= \frac{R_B}{kC_{Ao}^2(1 - X_A)^2}$$

The fixed annualized reactor cost from the Engineering Economics section in the NCEES Handbook is

$$A_F = V\left[P\left(A/P, i, n\right)\right]$$

$$= \frac{R_B}{kC_{Ao}^2(1 - X_A)^2}\left[P\left(A/P, i, n\right)\right]$$

The variable annual raw-material cost is

$$A_V = F_{Ao}U = \frac{R_B}{X_A}U$$

The total equivalent uniform annual cost, EUAC, is the fixed plus variable cost

$$A_T = A_F + A_V$$

$$= \frac{R_B}{kC_{Ao}^2(1 - X_A)^2}\left[P\left(A/P, i, n\right)\right] + \frac{R_B}{X_A}U$$

Minimize the total cost with respect to conversion. Take the derivative of the EUAC with respect to conversion and set it equal to zero,

$$\frac{dA_T}{dX_A} = 0 = \frac{d}{dX_A}\left(\begin{array}{c}\dfrac{R_B}{kC_{Ao}^2(1 - X_A)^2} \\ \times\left[P\left(A/P, i, n\right)\right] + \dfrac{R_B}{X_A}U\end{array}\right)$$

$$0 = \frac{R_B}{kC_{Ao}^2}\left[P\left(A/P, i, n\right)\right]\frac{d}{dX_A}\left(\frac{1}{(1 - X_A)^2}\right)$$

$$+ R_BU\left(\frac{d}{dX_A}\right)\left(\frac{1}{X_A}\right)$$

$$0 = \frac{R_B}{kC_{Ao}^2}\left[P\left(A/P, i, n\right)\right]\left(\frac{2}{(1 - X_A)^3}\right)$$

$$+ R_BU\left(-\frac{1}{X_A^2}\right)$$

Inserting numerical values, an equation in terms of optimal conversion results

$$0 = \frac{\left(0.031\,710 \dfrac{\text{mol}}{\text{s}}\right)\left(\dfrac{\$100,000}{1 \text{ m}^3}\right)}{\left(0.00001 \dfrac{\text{m}^3}{\text{s·mol}}\right)\left(1000 \dfrac{\text{mol}}{\text{m}^3}\right)^2}$$

$$\times\left[\frac{(0.1)(1+0.1)^{10}}{(1+0.1)^{10}-1}\text{yr}^{-1}\right]\left[\frac{2}{(1 - X_A)^3}\right]$$

$$+ \left(1\,000\,000 \frac{\text{mol}}{\text{yr}}\right)\left(\frac{\$0.10}{\text{mol}}\right)\left(-\frac{1}{X_A^2}\right)$$

This equation becomes,

$$\frac{103.21}{(1-X_A)^3} - \frac{100\,000}{X_A^2} = 0$$

Newton's method can be used to find the root from the Mathematics section of the NCEES Handbook,

$$f(X_A) = 0$$

$$X_{A,n+1} = X_{A,n} - \frac{fX_{A,n}}{f'X_{A,n}}$$

The first derivative of the function is

$$f'(X_{A,n}) = \frac{d}{dX_A}\left(\frac{103.21}{(1-X_A)^3} - \frac{100\,000}{X_A^2}\right)$$

$$= \frac{309.63}{(1-X_A)^4} + \frac{200\,000}{X_A^3}$$

Newton's iteration formula becomes

$$X_{A,n+1} = X_{A,n} - \frac{\dfrac{103.21}{(1-X_A)^3} - \dfrac{100\,000}{X_A^2}}{\dfrac{309.63}{(1-X_A)^4} + \dfrac{200\,000}{X_A^3}}$$

Iteration history can be seen as follows,

n	$X_{A,n}$
1	0.85
2	0.965 04
3	0.953 92
4	0.940 21
5	0.925 10
6	0.912 34
7	0.906 33
8	0.905 43
9	0.905 42

The optimal conversion is then

$$X_A = 0.90542 \times 100\%$$
$$= 90.542\% \quad (91\%)$$

The answer is C.

35. The reactor volume at the optimal conversion is

$$V = \frac{R_B}{kC_{Ao}^2(1-X_A)^2}$$

$$= \frac{0.031710 \ \dfrac{\text{mol}}{\text{s}}}{\left(0.00001 \ \dfrac{\text{m}^3}{\text{s·mol}}\right)\left(1000 \ \dfrac{\text{mol}}{\text{m}^3}\right)^2}$$
$$\times (1-0.90542)^2$$

$$= 0.35448 \text{ m}^3 \quad (0.4 \text{ m}^3)$$

The answer is B.

36. The EUAC at the optimal conversion is from the previous problem.

$$A_T = V\left[P(A/P,i,n)\right] + \frac{R_B}{X_A}U$$

$$= (0.35448 \text{ m}^3)\left(\frac{\$100,000}{1 \text{ m}^3}\right)$$

$$\times \left(\frac{(0.1)(1+0.1)^{10}}{(1+0.1)^{10}-1}\right)\frac{\$}{\text{yr}}$$

$$+ \left(\frac{1\,000\,000 \ \dfrac{\text{mol}}{\text{yr}}}{0.905\,42}\right)\left(\frac{\$0.10}{1 \text{ mol}}\right)$$

$$= \$116,210/\text{yr} \quad (\$100,000/\text{yr})$$

The answer is D.

37. For steady, incompressible flow, the mechanical energy (Bernoulli) balance equation for flow in conduits is given in the Fluid Mechanics section of the NCEES Handbook.

$$\frac{p_1}{\gamma} + z_1 + \frac{v_1^2}{2g} = \frac{p_2}{\gamma} + z_2 + \frac{v_2^2}{2g} + h_f + h_{f,\text{fitting}}$$

Note the restriction of the units of this equation to SI. If U.S. units are used, the equations including g_c must be used.

For pipes, the Darcy head-loss term is given in the Fluid Mechanics section of the NCEES Handbook [3].

$$h_f = f\left(\frac{L}{D}\right)\left(\frac{v^2}{2g}\right)$$

For fittings, the head-loss term is given in the Fluid Mechanics section of the NCEES Handbook.

$$h_{f,\text{fitting}} = C\frac{v^2}{2g}$$

For the current problem, let subscript 1 be at the surface of the liquid and subscript 2 be the pipe exit. The Bernoulli equation becomes

$$\frac{p_1}{\gamma} + z_1 + \frac{v_1^2}{2g} = \frac{p_2}{\gamma} + z_2 + \frac{v_2^2}{2g} + f\left(\frac{L}{D}\right)\left(\frac{v_2^2}{2g}\right)$$
$$+ \left(\begin{array}{c}C_{\text{entrance}} + C_{\text{elbow}}\\ + C_{\text{valve}} + C_{\text{exit}}\end{array}\right)\frac{v_2^2}{2g}$$

The tank and exit are both at atmospheric pressure. Since the tank has a large diameter, the surface is moving slowly.

$$p_1 = p_2$$
$$v_1 = 0$$

The Bernoulli equation becomes

$$z_1 = z_2 + \left(\begin{array}{c} 1 + f\dfrac{L}{D} + C_{\text{entrance}} \\ + C_{\text{elbow}} + C_{\text{valve}} + C_{\text{exit}} \end{array}\right) \dfrac{\text{v}_2^2}{2g}$$

Solving for velocity,

$$\text{v}_2 = \sqrt{\dfrac{2g(z_1 - z_2)}{1 + f\dfrac{L}{D} + C_{\text{entrance}} + C_{\text{elbow}} \\ + C_{\text{valve}} + C_{\text{exit}}}}$$

Inserting values from the Fluid Mechanics section in the NCEES Handbook,

$$\text{v}_2 = \sqrt{\dfrac{(2)\left(9.807 \, \dfrac{\text{m}}{\text{s}^2}\right)(9 \text{ m} - 5 \text{ m})}{1 + f\left(\dfrac{10 \text{ m}}{0.0525 \text{ m}}\right) + 0.5 + 0.9 + 0.2 + 1}}$$

$$= \sqrt{\dfrac{78.456}{3.6 + 190.48f} \dfrac{\text{m}}{\text{s}}}$$

Iterate on velocity, Reynolds number, and friction factor. To get the friction factor, the Reynolds number and properties of water are taken from the NCEES Handbook,

$$\text{Re} = \dfrac{\text{v}D\rho}{\mu} = \dfrac{\text{v}_2(0.0525 \text{ m})\left(999.1 \, \dfrac{\text{kg}}{\text{m}^3}\right)}{0.001\,139 \text{ Pa·s}}$$

$$= 46051.6V_2$$

For commercial steel pipe, from the Fluid Mechanics section of the NCEES Handbook [3] Moody Diagram, the relative roughness is

$$\dfrac{e}{D} = \dfrac{(0.046 \text{ mm})\left(\dfrac{1 \text{ m}}{1000 \text{ mm}}\right)}{0.0525 \text{ m}} \approx 0.0009$$

The friction factor can be obtained from the Fluid Mechanics section of the NCEES Handbook Moody Diagram or the Colebrook equation [7,8] (which, it can be seen, requires some iteration on f for solution).

$$\dfrac{1}{\sqrt{f}} = -2\log_{10}\left(\dfrac{\frac{e}{D}}{3.7} + \dfrac{2.512}{\text{Re}\sqrt{f}}\right) \quad [\text{Re} > 4000]$$

The iteration history is

n	v_2 (m/s)	Re = $4.6052 \times 10^4 V_2$	$f = f(\text{Re},(e/D))$	v_2(m/s) calculated
1	1.0	4.6052×10^4	0.024	3.099
2	3.099	1.4272×10^5	0.021	3.2130
3	3.2130	1.4797×10^5	0.021	3.2130

The final velocity is

$$\text{v}_2 = 3.2130 \text{ m/s} \quad (3.2 \text{ m/s})$$

The answer is B.

38. The volumetric flow rate, from the Fluid Mechanics section of the NCEES Handbook, is

$$Q = \text{v}A = \text{v}\dfrac{\pi}{4}D^2$$

$$= \left(3.2130 \, \dfrac{\text{m}}{\text{s}}\right)\left(\dfrac{\pi}{4}\right)(0.0525 \text{ m})^2\left(3600 \, \dfrac{\text{s}}{\text{h}}\right)$$

$$= 25.039 \text{ m}^3/\text{h} \quad (25 \text{ m}^3/\text{h})$$

The answer is D.

39. For the current problem, let subscript 1 be at the water surface of the source tank and subscript 2 be at the water surface of the receiving tank, with V the velocity in the pipe. The Bernoulli equation (Eq. 5.5) becomes (including the specific work—the work per unit mass—, w, and efficiency, η)

$$\dfrac{p_1}{\gamma} + z_1 + \dfrac{\text{v}_1^2}{2g} = \dfrac{p_2}{\gamma} + z_2 + \dfrac{\text{v}_2^2}{2g} + f\left(\dfrac{L}{D}\right)\left(\dfrac{\text{v}_2^2}{2g}\right)$$

$$+ \left(\begin{array}{c} C_{\text{entrance}} + C_{\text{elbow}} \\ + C_{\text{valve}} + C_{\text{exit}} \end{array}\right)$$

$$\times \left(\dfrac{\text{v}_2^2}{2g}\right) + \dfrac{\eta w}{g}$$

Note here the specific work is outward, or work done by the system on the surroundings, to be consistent with the Thermodynamics section of the NCEES Handbook. The tank and suction are both at atmospheric pressure. Since both tanks have a large diameter, the surfaces are moving slowly.

$$p_1 = p_2$$
$$\text{v}_1 \approx 0$$
$$\text{v}_2 \approx 0$$

The Bernoulli balance becomes

$$z_1 = z_2 + \left(\begin{array}{c} f\dfrac{L}{D} + C_{\text{entrance}} + C_{\text{elbow}} \\ + C_{\text{valve}} + C_{\text{exit}} \end{array}\right)$$

$$\times \dfrac{\text{v}_2^2}{2g} + \dfrac{\eta w}{g}$$

The average velocity in the pipe, from the Fluid Mechanics section of the NCEES Handbook, is

$$\text{v} = \dfrac{Q}{A} = \dfrac{Q}{\dfrac{\pi}{4}D^2}$$

$$= \dfrac{0.025 \, \dfrac{\text{m}^3}{\text{s}}}{\left(\dfrac{\pi}{4}\right)(0.0525 \text{ m})^2}$$

$$= 11.549 \text{ m/s}$$

The Reynolds number is

$$\text{Re} = \frac{\text{v}D\rho}{\mu}$$

$$= \frac{\left(11.549 \, \frac{\text{m}}{\text{s}}\right)(0.0525 \text{ m})\left(999.1 \, \frac{\text{kg}}{\text{m}^3}\right)}{0.001\,139 \text{ Pa·s}}$$

$$= 5.3185 \times 10^5$$

The friction factor can be obtained from the Moody (Stanton) Diagram in the Fluid Mechanics section of the NCEES Handbook.

$$f = 0.019$$

Solving for pump specific work,

$$w = \frac{1}{\eta}\left(\begin{array}{c} g\left(z_1 - z_2\right) - \left(\begin{array}{c} f\dfrac{L}{D} + C_{\text{entrance}} \\ + C_{\text{elbow}} + C_{\text{valve}} + C_{\text{exit}} \end{array}\right) \\ \times \dfrac{\text{v}^2}{2} \end{array}\right)$$

$$= \left(\frac{1}{0.70}\right)\left(\begin{array}{c} \left(9.807 \, \frac{\text{m}}{\text{s}^2}\right)(5 \text{ m} - 9 \text{ m}) \\ -\left(\begin{array}{c} (0.019)\left(\dfrac{10 \text{ m}}{0.0525 \text{ m}}\right) + 0.5 \\ +0.9 + 0.2 + 1.0 \end{array}\right) \\ \times \dfrac{\left(11.549 \, \frac{\text{m}}{\text{s}}\right)^2}{2} \end{array}\right)$$

$$= -648.53 \text{ J/kg}$$

To check this, note that the specific work is negative. This confirms that it is a pump (and not, for example, a turbine), since work on the system is being done by the surroundings. The mass flow rate, from the Fluid Mechanics section of the NCEES Handbook, is

$$\dot{m} = \rho \text{v} A = \rho \text{v} \frac{\pi}{4}D^2$$

$$= \left(999.1 \, \frac{\text{kg}}{\text{m}^3}\right)\left(11.549 \, \frac{\text{m}}{\text{s}}\right)\left(\frac{\pi}{4}\right)(0.0525 \text{ m})^2$$

$$= 24.978 \text{ kg/s}$$

The power required is the mass flow rate times the work,

$$\dot{W} = \dot{m}w$$

$$= \left(24.978 \, \frac{\text{kg}}{\text{s}}\right)\left(-648.53 \, \frac{\text{J}}{\text{kg}}\right)\left(\frac{1 \text{ kW}}{1000 \text{ W}}\right)$$

$$= -16.199 \text{ kW} \quad (-16 \text{ kW})$$

The answer is A.

40. The energy balance for open systems, neglecting potential and kinetic energy and assuming adiabatic conditions, is

$$\Delta H = Q = 0$$

For the enthalpy path from initial to final conditions, reactants can be brought from 500°C (773.15K) to 25°C (298.15K) and reacted to products at 25°C. Then the products can be brought from 25°C to the final unknown temperature.

$$\begin{aligned} \Delta H &= H_4 - H_1 \\ &= (H_2 - H_1) + (H_3 - H_2) + (H_4 - H_3) \\ &= Q \\ &= 0 \\ \Delta H &= \Delta H_R^o + \Delta H_r^o + \Delta H_P^o \\ &= Q \\ &= 0 \end{aligned}$$

The reactant enthalpy change with a basis of 1 mol of ethylene is

$$\begin{aligned} \Delta H_R^o &= \sum_{\text{reactants}} \nu_i \Delta \hat{H}_i^o \\ &= \int_{T_1}^{T_2}\left(\sum_{\text{reactants}} \nu_i c_{pi}^o\right) dT \\ &= \int_{773.15\text{K}}^{298.15\text{K}} (1 \text{ mol})\left(\begin{array}{c} 28.161 \, \dfrac{\text{J}}{\text{mol·K}} \\ + 0.054\,016T \, \dfrac{\text{J}}{\text{mol·K}} \end{array}\right) \\ &\quad \times dT \\ &= 28.161T + \left(\frac{0.054\,016}{2}\right)T^2 \Big|_{773.15\text{K}}^{298.15\text{K}} \\ &= -27\,120 \text{ J} \end{aligned}$$

Heat of reaction at 25°C is the weighted heats of formation of the products minus the reactants.

$$\begin{aligned} \Delta H_r^o &= \sum_{\text{products}} \nu_i \Delta \hat{H}_{fi}^o - \sum_{\text{reactants}} \nu_i \Delta \hat{H}_{fi}^o \\ &= (1 \text{ mol})\left(0 \, \frac{\text{J}}{\text{mol}}\right) + (1 \text{ mol})\left(-74\,520 \, \frac{\text{J}}{\text{mol}}\right) \\ &\quad - (1 \text{ mol})\left(52\,510 \, \frac{\text{J}}{\text{mol}}\right) \\ &= -127\,030 \text{ J} \end{aligned}$$

The product enthalpy change is

$$\Delta H_P^o = \sum_{\text{products}} \nu_i \Delta \hat{H}_i^o$$

$$= \int_{T_1}^{T_2} \left(\sum_{\text{products}} \nu_i c_{pi}^o \right) dT$$

$$= \int_{298.15\text{K}}^{T} \left(\begin{array}{l} (1\ \text{mol}) \\ \times \left(\begin{array}{l} 5.2173\ \dfrac{\text{J}}{\text{mol·K}} \\ + 0.011\,073T\ \dfrac{\text{J}}{\text{mol·K}} \end{array} \right) \\ + (1\ \text{mol}) \\ \times \left(\begin{array}{l} 22.192\ \dfrac{\text{J}}{\text{mol·K}} \\ + 0.043\,151T\ \dfrac{\text{J}}{\text{mol·K}} \end{array} \right) \end{array} \right)$$
$$\times\, dT$$

$$= 5.2173T\ \text{J} + \frac{0.011\,073}{2}T^2\ \text{J}$$
$$+ \left(22.192T\ \text{J} + \frac{0.043\,151}{2}T^2\ \text{J} \right)\Big|_{298.15\text{K}}^{T}$$
$$= \left(27.409T\ \text{J} + 0.027\,112T^2\ \text{J} \right)\Big|_{298.15\text{K}}^{T}$$
$$= 27.409T\ \text{J} + 0.027\,112T^2\ \text{J} - 10\,582\ \text{J}$$

The energy balance yields an expression for the product temperature.

$$\Delta H = \Delta H_R^o + \Delta H_r^o + \Delta H_P^o = Q$$
$$= 0$$
$$0 = -27\,120\ \text{J} - 127\,030\ \text{J}$$
$$+ \begin{array}{l} 27.409T\ \text{J} + 0.027\,112T^2\ \text{J} \\ -10582\ \text{J} \end{array}$$
$$0 = -164\,732\ \text{J} + 27.409T\ \text{J} + 0.027\,112T^2\ \text{J}$$

Solving for the product temperature and taking the positive root,

$$T = \frac{-27.409 + \sqrt{\begin{array}{l}(27.409)^2 - (4)(0.027\,112) \\ \times (-164\,732)\end{array}}}{(2)(0.027\,112)}$$
$$= 2010.8\text{K} \quad (2000\text{K})$$

The answer is C.

A more accurate calculation would include the high-pressure effect on the enthalpy.

41. Assume the ideal gas law holds,

$$PV = NRT$$

Ratioing final states over initial states,

$$\frac{P_2 V_2}{P_1 V_1} = \frac{N_2 R T_2}{N_1 R T_1}$$

If the solid carbon volume is neglected, the tank volume is constant and the gas moles are constant from the stoichiometry. The final pressure is

$$P_2 = P_1 \frac{V_1 N_2 R T_2}{V_2 N_1 R T_1}$$
$$= P_1 \frac{T_2}{T_1}$$
$$= (2000\ \text{bar}) \left(\frac{2010.8\text{K}}{773.15\text{K}} \right) \left(\frac{1\ \text{MPa}}{10\ \text{bar}} \right)$$
$$= 520.16\ \text{MPa} \quad (500\ \text{MPa})$$

The answer is C.

A more accurate calculation would include the high-pressure gas compressibility factor as well as the volume of the solid carbon.

42. The Reynolds number is

$$\text{Re} = \frac{DV\rho}{\mu} = \frac{(0.01\ \text{m})\left(1\ \dfrac{\text{m}}{\text{s}}\right)\left(1.2056\ \dfrac{\text{kg}}{\text{m}^3}\right)}{1.8 \times 10^{-5}\,\text{Pa·s}}$$
$$= 669.78$$

The Schmidt number is

$$\text{Sc} = \frac{\mu}{\rho D_m} = \frac{1.8 \times 10^{-5}\,\text{Pa·s}}{\left(1.2056\ \dfrac{\text{kg}}{\text{m}^3}\right)\left(6.2 \times 10^{-6}\dfrac{\text{m}^2}{\text{s}}\right)}$$
$$= 2.4081$$

The Sherwood number [11] is

$$\text{Sh} = 0.43 + (0.532)\,(\text{Re})^{0.5}\,(\text{Sc})^{0.31}$$
$$= 0.43 + (0.532)(669.78)^{0.5}(2.4081)^{0.31}$$
$$= 18.510$$

This is the case of napthalene (A) diffusing through stagnant air (B). Assume that partial pressure in the free air stream is zero at point 2, and the partial pressure is the vapor pressure at the surface of the cylinder at point 1.

$$p_{A1} = 6.67\ \text{Pa}$$
$$p_{A2} = 0\ \text{Pa}$$
$$P = 1\ \text{atm}$$
$$= 101\,325\ \text{Pa}$$

The log mean of the air partial pressure, Eq. 4.4, for A diffusing through stagnant B in gases is

$$(p_B)_{lm} = \frac{p_{B2} - p_{B1}}{\ln \frac{p_{B2}}{p_{B1}}}$$

$$= \frac{(P - p_{A2}) - (P - p_{A1})}{\ln \left(\frac{P - p_{A2}}{P - p_{A1}} \right)}$$

$$= \frac{(101\,325 \text{ Pa} - 0 \text{ Pa}) - (101\,325 \text{ Pa} - 6.67 \text{ Pa})}{\ln \left(\frac{101\,325 \text{ Pa} - 0 \text{ Pa}}{101\,325 \text{ Pa} - 6.67 \text{ Pa}} \right)}$$

$$= 101\,322 \text{ Pa}$$

The gas mass-transfer coefficient is

$$k_G = \left(\frac{Sh D_m}{(p_B)_{lm} D} \right) \left(\frac{P}{RT} \right)$$

$$= \frac{(18.510) \left(6.2 \times 10^{-6} \frac{m^2}{s} \right)}{(101\,322 \text{ Pa})(0.01 \text{ m})}$$

$$\times \left(\frac{101\,325 \text{ Pa}}{\left(8.314 \frac{Pa \cdot m^3}{mol \cdot K} \right) (20°C + 273.15°)} \right)$$

$$= 4.7088 \times 10^{-6} \text{ mol/m}^2 \cdot s \cdot Pa$$

$$(5 \times 10^{-6} \text{ mol/m}^2 \cdot s \cdot Pa)$$

The answer is A.

43. The molar flux using an individual mass transfer coefficient for A diffusing through stagnant B in gases is

$$\frac{N_A}{A} = k_G (p_{A1} - p_{A2})$$

$$= \left(4.7088 \times 10^{-6} \frac{mol}{m^2 \cdot s \cdot Pa} \right) (6.67 \text{ Pa} - 0 \text{ Pa})$$

$$= 3.1408 \times 10^{-5} \text{ mol/m}^2 \cdot s$$

$$(3 \times 10^{-5} \text{ mol/m}^2 \cdot s)$$

The answer is A.

44. The compound with the largest solubility-product constant, K_{SP}, will dissolve the easiest. In this case, the answer is magnesium hydroxide, $Mg(OH)_2$.

The answer is C.

45. pH is estimated from the following expression.

$$pH = -\log[H^+]$$

$$= -\log \left[5.0 \times 10^{-6} \frac{mol}{L} \right]$$

$$= 5.3$$

The answer is B.

46. The equilibrium constant can be expressed as

$$K_{eq} = \frac{[H^+][ClO^-]}{HClO}$$

Substituting,

$$3.2 \times 10^{-8} = \frac{[H^+][1.5 \times 10^{-5}]}{[0.7 \times 10^{-7}]}$$

Rearrange and solve for $[H^+]$ to obtain,

$$[H^+] = 1.5 \times 10^{-10}$$

The answer is A.

47. Butyl refers to four carbons: 1) methyl group attached at the second of four carbons (isobutyl alcohol); 2) methyl group attached at the third of four carbons (sec-butyl alcohol); 3) methyl group attached at the second of three carbons (isopropyl alcohol), and 4) methyl alcohol.

The answer is B.

48. The equation given in option D is not balanced. Changing O_2 from 12 mol to 15 mol would balance the equation.

The answer is D.

49. The compound shown represents p-xylene. The other compounds mentioned would be represented as follows.

toluene

o-xylene

m-xylene

The answer is D.

50. Calculate the component mass flow rates in the feed.

$$\dot{m}_{\text{propane},i} = x_{\text{propane}}\dot{m}_{\text{feed}}$$
$$= (0.20)\left(1500\ \frac{\text{kg}}{\text{g}}\right)$$
$$= 300\ \text{kg/h}$$

$$\dot{m}_{\text{propylene},i} = x_{\text{propylene}}\dot{m}_{\text{feed}}$$
$$= (0.25)\left(1500\ \frac{\text{kg}}{\text{h}}\right)$$
$$= 375\ \text{kg/h}$$

$$\dot{m}_{\text{butane},i} = x_{\text{butane}}\dot{m}_{\text{feed}}$$
$$= (0.35)\left(1500\ \frac{\text{kg}}{\text{h}}\right)$$
$$= 525\ \text{kg/h}$$

$$\dot{m}_{\text{pentane},i} = x_{\text{pentane}}\dot{m}_{\text{feed}}$$
$$= (0.20)\left(1500\ \frac{\text{kg}}{\text{h}}\right)$$
$$= 300\ \text{kg/h}$$

Using the stated propylene recovery value of 95%, calculate the propylene mass flow rates in the overhead and bottoms product streams.

$$\dot{m}_{\text{propylene},o} = 0.95\dot{m}_{\text{propylene},i}$$
$$(0.95)\left(375\ \frac{\text{kg}}{\text{h}}\right)$$
$$= 356.3\ \text{kg/h}$$

$$\dot{m}_{\text{propylene},B} = \dot{m}_{\text{propylene},i} - \dot{m}_{\text{propylene},o}$$
$$= 375\ \frac{\text{kg}}{\text{h}} - 356.3\ \frac{\text{kg}}{\text{h}}$$
$$= 18.7\ \text{kg/h}$$

Therefore, the total mass flow rate of the bottoms stream is

$$\dot{m}_{\text{total},B} = \frac{\dot{m}_{\text{propylene},B}}{x_{\text{propylene},B}}$$

The percent propylene in the bottoms stream is given in the problem statement as 5%. Its mass fraction is therefore 0.05.

$$= \frac{18.7\ \dfrac{\text{kg}}{\text{h}}}{0.05}$$
$$= 374\ \text{kg/h}\quad (370\ \text{kg/h})$$

The answer is B.

51. Determine the overhead mass flow rate of propylene and the total overhead stream mass flow rate. Using these values calculate the percent propylene in the overhead using the following equations.

$$\dot{m}_o = \dot{m}_F - \dot{m}_B$$
$$= 1500\ \frac{\text{kg}}{\text{h}} - 1000\ \frac{\text{kg}}{\text{h}}$$
$$= 500\ \text{kg/h}$$

$$\dot{m}_{\text{propylene},F} = \dot{m}_F x_{\text{propylene},F}$$
$$= \left(1500\ \frac{\text{kg}}{\text{h}}\right)(0.25)$$
$$= 375\ \text{kg/h}$$

$$\dot{m}_{\text{propylene},B} = \dot{m}_B x_{\text{propylene},B}$$
$$= \left(1000\ \frac{\text{kg}}{\text{h}}\right)(0.05)$$
$$= 50\ \text{kg/h}$$

$$\dot{m}_{\text{propylene},o} = \dot{m}_{\text{propylene},F} - \dot{m}_{\text{propylene},B}$$
$$= 375\ \frac{\text{kg}}{\text{h}} - 50\ \frac{\text{kg}}{\text{h}}$$
$$= 325\ \text{kg/h}$$

$$\%\text{propylene},o = \frac{\dot{m}_{\text{propylene},o}}{\dot{m}_o} \times 100\%$$
$$= \left(\frac{325\ \dfrac{\text{kg}}{\text{h}}}{500\ \dfrac{\text{kg}}{\text{h}}}\right) \times 100\%$$
$$= 65.0\%$$

The answer is B.

52. Calculate the heat content, Q, of each stream using the following formulas and the specific heat information, h, provided in the table.

$$\dot{m}_1 = 3000 \text{ kg/h}$$

$$\dot{m}_6 = \dot{m}_1 - \dot{m}_7$$
$$= 3000 \,\frac{\text{kg}}{\text{h}} - 800 \,\frac{\text{kg}}{\text{h}}$$
$$= 2200 \text{ kg/h}$$

$$h_6 = h_3 = 60 \text{ kcal/kg}$$

$$Q_6 = \dot{m}_6 h_6$$
$$= \left(2200 \,\frac{\text{kg}}{\text{h}} \right) \left(60 \,\frac{\text{kcal}}{\text{kg}} \right)$$
$$= 132\,000 \text{ kcal/h}$$

$$\dot{m}_7 = 800 \text{ kg/h}$$

$$h_7 = h_5 = 100 \text{ kcal/kg}$$

$$Q_7 = (\dot{m}_7)(h_7)$$
$$= \left(800 \,\frac{\text{kg}}{\text{h}} \right) \left(100 \,\frac{\text{kcal}}{\text{kg}} \right)$$
$$= 80\,000 \text{ kcal/h}$$

$$Q_1 = Q_6 + Q_7$$
$$= 132\,000 \,\frac{\text{kcal}}{\text{h}} + 80\,000 \,\frac{\text{kcal}}{\text{h}}$$
$$= 212\,000 \text{ kcal/h}$$

$$h_1 = \frac{Q_1}{\dot{m}_1}$$
$$= \frac{212\,000 \,\dfrac{\text{kcal}}{\text{h}}}{3000 \,\dfrac{\text{kg}}{\text{h}}}$$
$$= 70.6 \text{ kcal/kg} \quad (71 \text{ kcal/kg})$$

The answer is A.

53. Determine the overhead mass flow rate of propylene and the total overhead stream mass flow rate. Using these values calculate the percent propylene in the overhead using the following equation.

$$\dot{m}_o = \dot{m}_F - \dot{m}_B$$
$$= 1500 \,\frac{\text{kg}}{\text{h}} - 500 \,\frac{\text{kg}}{\text{h}}$$
$$= 1000 \,\frac{\text{kg}}{\text{h}}$$

$$\dot{m}_{\text{propylene},F} = \dot{m}_F x_{\text{propylene},F}$$
$$= \left(1500 \,\frac{\text{kg}}{\text{h}} \right)(0.25)$$
$$= 375 \text{ kg/h}$$

$$\dot{m}_{\text{propylene},B} = \dot{m}_B x_{\text{propylene},B}$$
$$= \left(500 \,\frac{\text{kg}}{\text{h}} \right)(0.05)$$
$$= 25 \text{ kg/h}$$

$$\dot{m}_{\text{propylene},o} = \dot{m}_{\text{propylene},F} - \dot{m}_{\text{propylene},B}$$
$$= 375 \,\frac{\text{kg}}{\text{h}} - 25 \,\frac{\text{kg}}{\text{h}}$$
$$= 350 \text{ kg/h}$$

$$\%\text{propylene},o = \frac{\dot{m}_{\text{propylene},o}}{\dot{m}_o} \times 100\%$$
$$= \left(\frac{350 \,\dfrac{\text{kg}}{\text{h}}}{1000 \,\dfrac{\text{kg}}{\text{h}}} \right) \times 100\%$$
$$= 35\%$$

The answer is D.

54. According to the affinity law,

$$\frac{Q_2}{Q_1} = \frac{D_2}{D_1}$$

Convert the 200 mm and 230 mm impeller diameters to meters, and then rearrange to solve for the new volumetric flow rate after the impeller size change.

$$Q_2 = \frac{Q_1 D_2}{D_1}$$
$$= \frac{\left(250 \,\dfrac{\text{m}^3}{\text{h}} \right)(0.230 \text{ m})}{0.200 \text{ m}}$$
$$= 287 \text{ m}^3/\text{h} \quad (290 \text{ m}^3/\text{h})$$

The answer is B.

55. Determine the ratio of the branch 3 volumetric flow rate to the branch 2 volumetric flow rate ignoring the pipe friction.

$$\frac{Q_3}{Q_2} = \sqrt{ \left(\frac{L_2}{L_3} \right) \left(\frac{D_3}{D_2} \right)^5 }$$
$$= \sqrt{ \left(\frac{1 \text{ m}}{1.5 \text{ m}} \right) \left(\frac{0.150 \text{ m}}{0.100 \text{ m}} \right)^5 }$$
$$= 2.25$$

Therefore, the branch 3 volumetric flow rate is

$$Q_3 = 2.25 Q_2$$

Given that the length of branch 3 is 50% more than the length of branch 2,

$$Q_3 = Q_1 - Q_2$$

Substituting,

$$2.25 Q_2 = Q_1 - Q_2$$

Rearranging,

$$2.25Q_2 + Q_2 = Q_1$$
$$3.25Q_2 = Q_1$$

Rearranging to solve for the branch 2 volumetric flow rate,

$$Q_2 = \frac{Q_1}{3.25}$$
$$= \frac{1000 \ \frac{m^3}{h}}{3.25}$$
$$= 307.7 \ m^3/h \quad (310 \ m^3/h)$$

The answer is B.

56. Determine the ratio of the branch 3 volumetric flow rate to the branch 2 volumetric flow rate. Pipe friction cannot be ignored.

$$\frac{Q_3}{Q_2} = \sqrt{\left(\frac{f_2}{f_3}\right)\left(\frac{L_2}{L_3}\right)\left(\frac{D_3}{D_2}\right)}$$
$$= \sqrt{\left(\frac{0.015}{0.025}\right)\left(\frac{1 \ m}{1.5 \ m}\right)\left(\frac{0.150 \ m}{0.100 \ m}\right)^5}$$
$$= 1.74$$

Therefore, the branch 3 volumetric flow rate is

$$Q_3 = 1.74Q_2$$

The volumetric flow rate in branch 3 is equal to the volumetric flow rate in branch 1 minus the volumetric flow rate in branch 2.

$$Q_3 = Q_1 - Q_2$$

Substituting,

$$1.74Q_2 = Q_1 - Q_2$$
$$1.74Q_2 + Q_2 = Q_1$$
$$2.74Q_2 = Q_1$$

Rearranging to solve for the branch 2 volumetric flow rate,

$$Q_2 = \frac{Q_1}{2.74}$$
$$= \frac{1000 \ \frac{m^3}{h}}{2.74}$$
$$= 365 \ m^3/h \quad (370 \ m^3/h)$$

The answer is C.

57. Along with the items described—a simplified sketch of the system, a mass balance, pressures, and temperatures—a process flow diagram also contains major control elements.

The answer is C.

58. The new cost can be determined from

$$C_2 = C_1 \frac{Cap_2}{Cap_1}$$

The cost of the exchangers is

$$C_{2,\text{exchangers}} = (\$10 \ M)\left(\frac{50 \ \frac{tons}{yr}}{25 \ \frac{tons}{yr}}\right)^{0.55}$$
$$= \$14.6 \ M$$

The cost of the vessels is

$$C_{2,\text{vessels}} = (\$15 \ M)\left(\frac{50 \ \frac{tons}{yr}}{25 \ \frac{tons}{yr}}\right)^{0.50}$$
$$= \$21.2 \ M$$

The cost of the pumps is

$$C_{2,\text{pumps}} = (\$10 \ M)\left(\frac{50 \ \frac{tons}{yr}}{25 \ \frac{tons}{yr}}\right)^{0.60}$$
$$= \$15.2 \ M$$

The cost of the piping is

$$C_{2,\text{piping}} = (\$15 \ M)\left(\frac{50 \ \frac{tons}{yr}}{25 \ \frac{tons}{yr}}\right)^{0.70}$$
$$= \$24.4 \ M$$

The cost of the miscellaneous equipment is

$$C_{2,\text{misc}} = (\$5 \ M)\left(\frac{50 \ \frac{tons}{yr}}{25 \ \frac{tons}{yr}}\right)^{0.75}$$
$$= \$8.4 \ M$$

Sum these costs to obtain the total new cost.

$$C_{2,\text{total}} = \Sigma C_2$$
$$= \$14.6 \ M + \$21.2 \ M + \$15.2 \ M$$
$$+ \$24.4 \ M + \$8.4 \ M$$
$$= \$83.8 \ M \quad (\$84 \ M)$$

The answer is B.

59. Exponentiation is performed before multiplication, division, or addition.

The answer is C.

60. The equation for standard deviation is

$$\sigma = \sqrt{\frac{\sum(X - \overline{X})^2}{n - 1}}$$

First, sum the additive levels to determine the total additive level.

$$\bar{X}_{\text{total}} = \Sigma \bar{X}$$

$$= \frac{\begin{array}{c} 2300 \text{ ppm} + 2800 \text{ ppm} + 2500 \text{ ppm} \\ + 2600 \text{ ppm} + 3100 \text{ ppm} \end{array}}{5}$$

$$= 2660 \text{ ppm}$$

Determine the sums of the additive levels, as in the following table.

calculation	$X - \overline{X}$ (ppm)	$(X - \overline{X})^2$ (ppm)
2300 ppm − 2660 ppm =	−360	129 600
2800 ppm − 2660 ppm =	140	19 600
2500 ppm − 2660 ppm =	−160	25 600
2600 ppm − 2660 ppm =	−60	3600
3100 ppm − 2660 ppm =	440	193 600
sum =	0	372 000

Substitute and solve for the standard deviation.

$$\sigma = \sqrt{\frac{372\,000 \text{ ppm}}{4}}$$

$$= 305 \text{ ppm} \quad (310 \text{ ppm})$$

The answer is B.

Where Can I Get an Online Practice Exam?

A realistic full-length practice exam is offered by **feprep.com**. The online *Practice Exam* environment at **feprep.com** accurately simulates the official computer-based testing (CBT) experience. It uses a graphical user interface that is equivalent to what is used during the actual exam. Important onscreen features include

- side-by-side presentation of questions and reference material suitable for 24-in monitors (with resizing option for smaller monitors)
- a fully searchable set of FE equations, tables, and figures equivalent to the NCEES *FE Reference Handbook*
- exam-like navigation (answer, skip, next, previous, flag for review, etc.)
- a timer, to simulate the exam's two sessions and break period
- a summary of all selected answers to review prior to submitting for grading

In addition, unlike the actual FE exam, the *Practice Exam* environment at **feprep.com** offers

- the ability to pause the examination for convenience
- immediate grading
- reporting of performance by knowledge area
- access to complete solutions for all problems

An Actual Screenshot of the FEPrep Exam Simulator

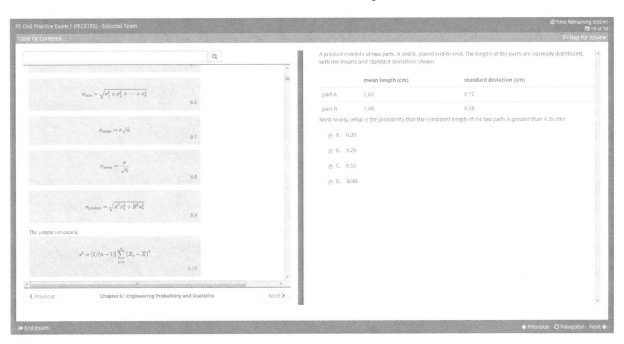